NORTH AMERICAN CONTINENTAL MARGINS

A Synthesis and Planning Workshop

Report of the
North American Continental Margins Working Group for
the U.S. Carbon Cycle Scientific Steering Group
and Interagency Working Group

U.S. Carbon Cycle Science Program
Washington D.C.

Editors
Burke Hales, Wei-Jun Cai, B. Greg Mitchell,
Christopher L. Sabine, and Oscar Schofield

The workshop was supported by the Carbon Cycle Interagency Working Gloup (CCIWG). The CCIWG includes the Department of Energy, the U. S. Geological Survey of the U. S. Department of the Interior, the National Aeronautics and Space Administration, the National Institute of Standards and Technology, the National Oceanic and Atmospheric Administration of the U. S. Department of Commerce, the National Science Foundation, and the U. S. Department of Agriculture. This report was printed by the University Corporation for Atmospheric Research (UCAR) under award number NA06OAR4310119 from the National Oceanic and Atmospheric Administration, U. S. Department of Commerce. The statements, findings, conclusions, and recomendations are those of the authors and do not necessarily reflect the views of any agency or program.

Recommended citation:
Burke Hales, Wei-Jun Cai, B. Greg Mitchell, Christopher L. Sabine, and Oscar Schofield [eds.], 2008: *North American Continental Margins: A Synthesis and Planning Workshop*. Report of the North American Continental Margins Working Group for the U.S. Carbon Cycle Scientific Steering Group and Interagency Working Group. U.S. Carbon Cycle Science Program, Washington, DC, 110 pp.

Table of Contents

Executive Summary

Continental margins represent a potentially large, but largely unconstrained, flux of CO_2 between the coastal ocean surface and atmosphere, with efforts to predict this flux generating estimates of either a sink or a source of approximately 1 Pg C yr^{-1}, which is significant globally relative to, e.g., pelagic air-sea exchange. Large variability inherent in these settings makes extrapolation based on sparse sampling difficult. Further confounding these efforts are uncertainties as to the boundaries of the coastal margins, both in seaward and landward extent. There is sensitivity of global margin air-sea CO_2-flux calculations to the inclusion or exclusion of estuaries. Other marginal environments, such as salt-marshes, mangrove forests, and tide flats may exchange large quantities of CO_2 directly with the atmosphere, and it is unclear whether these fluxes are accounted for in regional to global scale CO_2 budgets. In addition to air-sea gas exchange, the modes and magnitudes of carbon transport across other key boundaries in the coastal oceans are poorly understood. Net exchange between pelagic and coastal oceans; between estuaries and coastal oceans; between the coastal water column and seafloor; and between subaerial coastal environments and the atmosphere are all areas that have received limited study.

The knowledge of carbon cycling in North America's continental margins suffers similar deficiencies. There have been few integrative, carbon-focused, interdisciplinary studies in these settings, and only very recently has the totality of surface pCO$_2$ data in the waters surrounding North America been comprehensively evaluated. This effort suggests that the net air-sea flux in the open coastal waters (i.e., excluding estuaries and embayments) is nearly zero, but with uncertainty approaching ±40 Tg C yr^{-1}. The small net flux estimate appears to be a sensitive balance between regions of large CO_2 uptake at high latitudes and large CO_2 offgassing at low latitudes. These regions are the most undersampled of the coastal oceans, and raise the possibility of even larger uncertainties. North America's continental margins are defined by large variability in carbon-cycle-relevant processes and parameters. There is large regional variation in the importance of processes that control carbon transfer between terrestrial, oceanic, and atmospheric reservoirs,

and study of the carbon cycling in the entirety of these margins is complicated as a result.

In September of 2005, **North American Continental Margins: A Synthesis and Planning Workshop** was held in Boulder, Colorado to assess the state of carbon cycle science in the margins surrounding North America, and to offer recommendations guiding future research. The meeting was attended by over 50 scientists representing over 40 institutions and agencies, with expertise relevant to the measurement and interpretation of carbon cycling in the major marginal settings surrounding North America. The dominant processes driving carbon transport and transformation in the major geographical regions (Pacific, Atlantic, Gulf, and Arctic coasts) of the margins, in addition to the Laurentian Great Lakes and river/estuary systems, were discussed, as well as subregional variations in this process-dominance. State-of-the-art in-water and remote-sensing measurement approaches were reviewed with respect to their application in the coastal setting, as were synthetic approaches such as biogeochemical modeling based on general circulation and box-models, and data assimilation. Further discussion involved the needs for clear process-based definition of geographical subregions, integrative results of process study, and the relevance of historical datasets that may not have been applied to carbon cycle interpretations.

The group made several recommendations. These were:

- Increase the platform-of-opportunity measurements in the coastal settings whenever possible. Specifically, take advantage of existing moored deployment opportunities (such as navigational and NOAA NDBC buoys), and of ships that operate primarily in coastal settings (including coastal research and commercial/recreational vessels). This may require the development of new technology that is better suited for these deployments.

- Encourage examination of existing data that may reside in the gray literature or in data reports from non carbon-focused research programs. There was a general consensus that there is much data that has been undersynthesized in the context of carbon cycling. Nutrient or oxygen measurements; biological standing stock data; and hydrographic or circulation observations among others could all be interpreted from the perspective of the carbon cycle.

- Develop models that are better suited for application to coastal carbon cycling, both in terms of appropriate spatial and temporal resolution, and also in terms of inclusion of processes not typically incorporated in oceanic models.

- Improve remote-sensing algorithms so that complications in nearshore waters can be overcome. These include increasing spatial and temporal resolution, improving atmospheric corrections, and accounting for complicating factors such as increased sediment load and CDOM concentrations.

- Develop new in-water technologies for increased coverage of carbon-relevant parameters. Chemical analysis and sensing systems may need to become smaller and more robust to fit on coastal deployments-of-opportunity, which may not have the space or support infrastructure that exists on dedicated oceanographic research platforms. They may also need to be more affordable as high spatial and temporal resolution coverage drives the required number of deployed systems up. Some of these improvements may be facilitated by relaxed demands for precision in these highly variable systems.

- Define the continental margin as the environment contained between the head of tide and the exclusive economic zone, in addition to the Laurentian Great Lakes. Objectively divide the margins into a finite set of subregions within which focused process studies are representative of the entire subregion.

- Devise field measurement approaches for determining *net* fluxes of *total* carbon across key regime boundaries, such as the estuary-ocean and coastal-pelagic boundaries, in addition to the air-sea interface.

- Develop a plan for integrated process and synthesis study in representative subregions previously identified by the efforts listed above. Base this plan on a control-volume concept, where net reactions of carbon within, and net fluxes across key boundaries of, the control volume can be constrained. Encourage parallel implementation of modeling and in-water process study, as opposed to modeling studies that follow fieldwork.

The group did not make recommendations for strategies or timelines for agency implementation of these recommendations. Some of the above recommendations could obviously be implemented immediately through the existing funding structure. There is clear benefit in some of these efforts preceding others—for example, the synthesis of existing datasets and the definition of study subregions would help in planning the process and synthesis studies—while others—such as the development of improved models, remote sensing algorithms, and new measurement technologies—could proceed in parallel.

Introduction and Background

Burke Hales
College of Oceanic and Atmospheric Sciences
Oregon State University

Wei-Jun Cai
University of Georgia

B. Greg Mitchell
Scripps Institution of Oceanography
University of California, San Diego

Chris Sabine
Pacific Marine Environmental Laboratory
National Oceanographic and Atmospheric Administration

Oscar Schofield
Institute of Marine and Coastal Sciences
Rutgers University

Continental margins and the global carbon cycle

The significance of coastal oceans to global exchange of CO_2 with the atmosphere is now receiving significant attention (Bianchi, et al. 2005; Borges 2005; Cai et al. 2006; Cai et al. 2003a; Ducklow and McAllister, 2005; Chen et al. 2004; DeGrandpre et al. 2002; Frankignoulle and Borges, 2001; Friederich et al. 2002; Hales et al. 2005; Ianson et al. 2003; Ianson and Allen, 2002; Smith and Hollibaugh, 1993; Thomas et al. 2004; Tsunogai et al. 1999; Yool and Fasham, 2001). In all of these studies, the authors noted either large deviations of surface water pCO_2 from atmospheric saturation, or large imbalances in macroscopic carbon budgets. These were used to argue for significant global coastal fluxes despite the small area of coastal oceans. Many of these studies focused on large coastal uptake of atmospheric CO_2, but others (notably Smith and Hollibaugh, 1993, and Cai et al. 2003a) showed that coastal areas might act as strong CO_2 sources as well.

Impacts on estimates of surface-atmospheric CO_2 exchange

One method for estimation of net regional land-atmosphere CO_2 fluxes is the atmospheric inversion approach. This relies on a sparse flask-sampling network of surface and aircraft-based atmospheric CO_2 values, which, in conjunction with open-ocean surface CO_2 measurements and models of atmospheric transport and terrestrial ecosystems, can be used to constrain regional fluxes of CO_2 to and from the atmosphere (see, e.g., Baker et al. 2006). Since the open ocean

climatologies (e.g., Takahashi et al. 2002) do not include ocean margins, net air-sea CO_2 exchange from the coastal oceans will be attributed to other regions in the inversion method. This effect may be significant. Lueker et al. (2003) showed that fluxes from the coastal ocean can have a large impact on the chemistry of airmasses flowing onto the North American continent. If a net air-sea flux of ± 20 mmol CO_2 m^{-2} d^{-1}, as is commonly seen in the coastal ocean, could change the pCO_2 of a 500 m column of marine-boundary layer air for a day, that airmass would change its pCO_2 by ±1 µatm (Chavez et al. 2007). This is about an order of magnitude greater than the analytical precision of the flask-network measurements, and represents a significant uncertainty in the flux estimated by the inverse method.

This issue is particularly important in regions where mean airflow is cross-shore. For example, mean circulation patterns push air from the open Arctic and North Pacific-oceans onto the North American continent at mid- to high latitudes throughout much of the year, while summer air movements result in a mean onshore flow of marine air across the coasts of the Gulf of Mexico and the South Atlantic Bight (Figure 1.1). Consistent offshore flow occurs across the mid-north Atlantic coasts (Figure 1.1). These are all areas where large area-specific air-sea CO_2 fluxes have been previously documented (Hales et al. 2005; Cai et al. 2003a; Lohrenz and Cai, 2006; Bates et al. 2006; Degrandpre et al. 2002; Vandemark et al. 2006). While the inversion methods will probably capture total fluxes in any case, the ignorance of large fluxes at regional boundaries may result in assignment of coastal

3

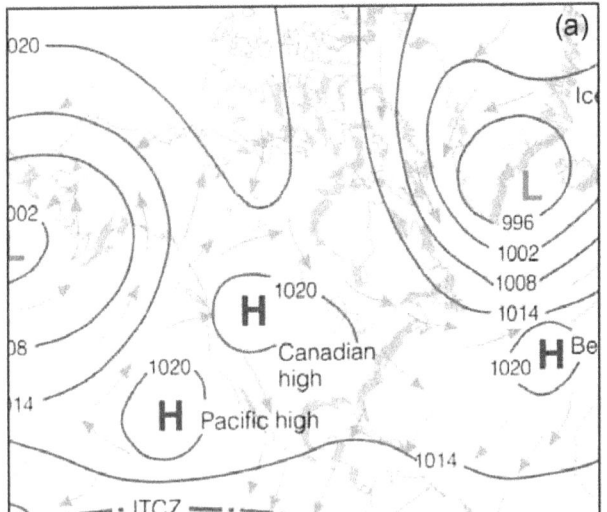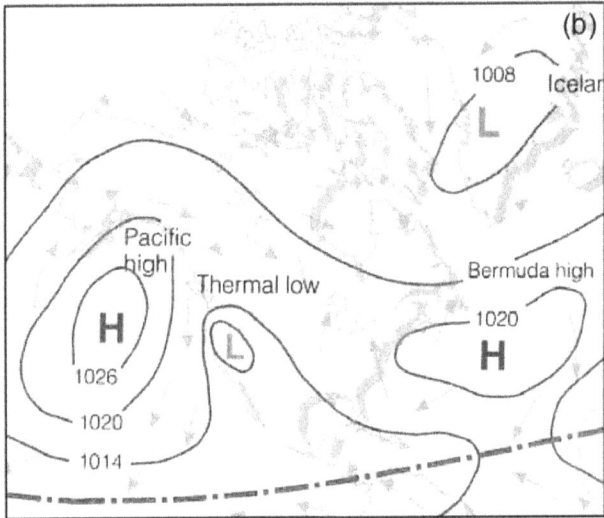

Figure 1.1. Mean atmospheric transport (arrows) across the North American coasts in a) January and b) July. Modified from Ahrens (2007).

fluxes to the terrestrial realm, or of coastal fluxes to the open ocean. Flux estimates from inversions are most reliable at the largest of scales, and the bias introduced by ignoring potentially large fluxes at continental margins is unlikely to affect estimates of hemispheric and global totals. At regional and continental scales, however, this bias could be more significant. Inversions generally impose an assumed spatial distribution on surface fluxes, and since those patterns do not have a representation of fluxes at continental margins, coastal fluxes will be attributed either to the open ocean or to terrestrial processes.

Undersampling of coastal margin carbon cycling

Despite the potential importance of the coastal oceans to the global atmospheric CO_2 budget, the continental shelves are greatly undersampled and beg for a more extensive observation network. The tendency of coastal ocean researchers has been to extrapolate their time- and space-limited results over large areas, and this may be tenuous given the short temporal and spatial scales of variability in the coastal ocean. The large observed ranges in coastal ocean surface pCO_2 exceed the dynamic range observed in the open ocean, and can occur at spatial scales as small as a few kilometers and over temporal scales of several hours (see, e.g., Cai et al. 2003a; Friederich et al. 2002; Hales et al. 2005). Further confounding the issue is the documentation of longer temperal and spatial-scale variability, with clear

impacts of interannual cycles (e.g., Chavez et al. 2002; Peterson and Schwing, 2003) on carbon-cycle relevant processes in the coastal ocean, and clear alongshore differences within given coastal regions (compare, e.g., Cai et al. 2003a and DeGrandpre et al. 2002).

To provide an initial estimate of global air-sea CO_2 flux on continental shelves, one must use a suitable classification that accounts for differences in ocean circulation, morphology, latitude, etc., and is also consistent with currently available data on shelf CO_2 fluxes. Two recent studies have attempted this. Borges (2005) (Figure 1.2) compiled existing coastal and estuarine air-sea CO_2 flux studies, and made a global coastal air-sea flux estimate of an uptake of 0.4 Pg C yr^{-1}. Borges et al. (2005, 2006) found, however, that when estuaries and embayments were included the total coastal air-sea flux was nearly negated by an estimated offgassing of 0.4 Pg C yr^{-1} from high-pCO_2 estuarine waters. Cai et al. (2006) made an effort to simplify continental shelves into three major types and seven provinces, thus limiting the unavoidable extrapolation of sparse data only to regions represented by individual datasets. This exercise yielded global shelf air to sea CO_2 flux of 0.2 ± 0.2 Pg C yr^{-1}. Despite these efforts, huge regions of the global continental margins remain largely unsampled, and best-effort extrapolations only stand to be improved by further study.

The importance of continental margins to global carbon cycling is not limited to air-sea CO_2 exchange alone. The advent of ocean color remote sensing

Figure 1.2. Two examples of databases used to globally extrapolate coastal annual air-sea CO$_2$ flux measurements. Compilations from a) Cai et al. (2006), and b) Borges (2005).

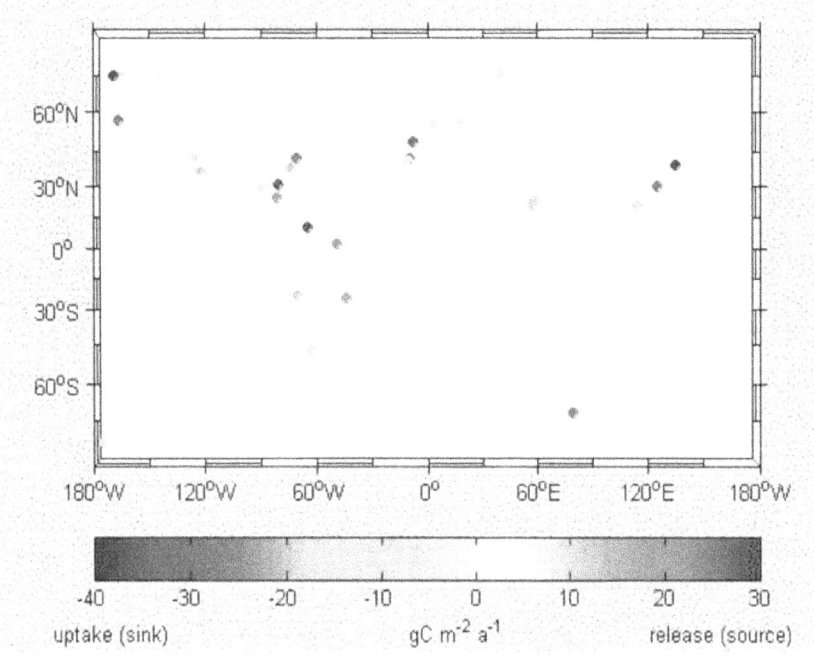

(a)

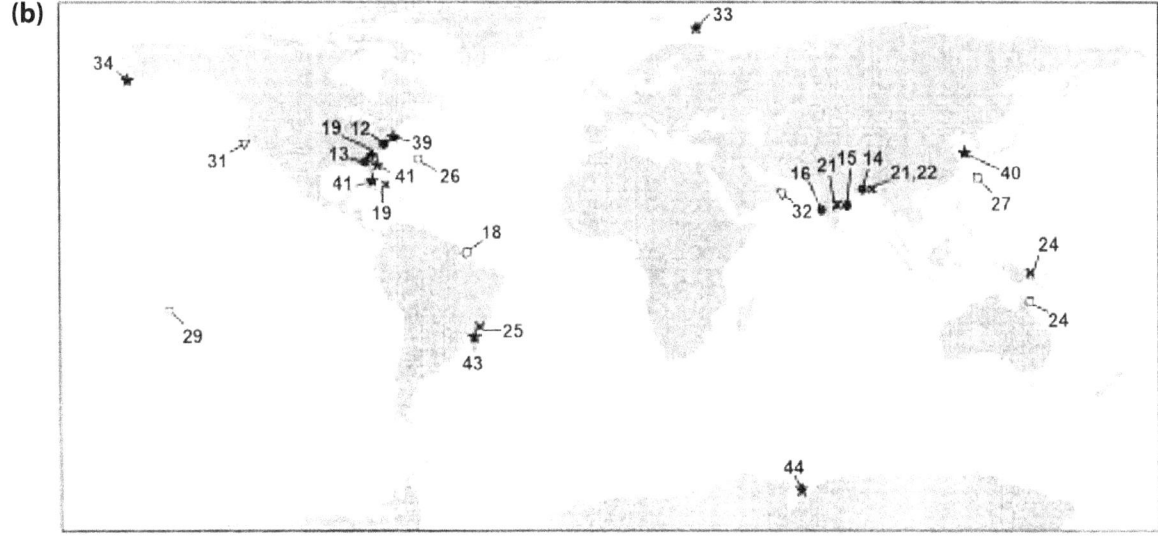

(b)

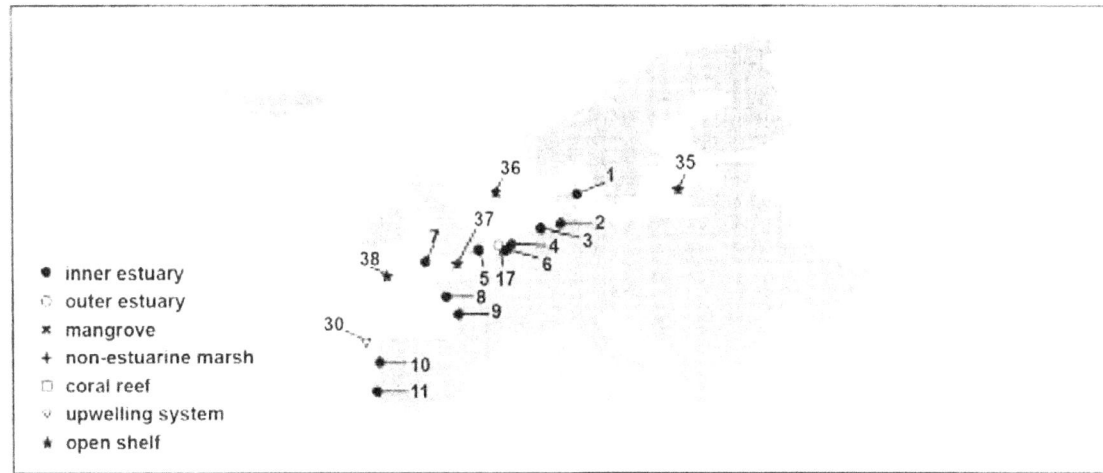

allowed the elevated biomass of the world's continental shelves to be observed (Figure 1.3). Along with elevated biomass, primary and net productivity rates are high in the coastal oceans. Algorithms for prediction of primary production from ocean color show intensification of primary productivity at the coastlines fueled by the nutrients provided by upwelling, rivers, and the atmosphere. Also associated with the high particulate carbon are high concentrations of terrestrial and marine-origin dissolved organic carbon. The fate of this carbon varies with the continental margin and thus overall shelf morphology significantly impacts the degree with which carbon produced on the shelf is exported to the continental slopes. Regardless, current export productivity estimates suggest that, globally, coastal oceans account for nearly a third of global ocean primary productivity and half the ocean export productivity (Muller-Karger et al. 2005).

Continental margins are the sites where exchange between terrestrial and pelagic ecosystems is determined. Rivers deliver about 0.45 Pg C yr[-1] of terrestrial POC and DOC to estuaries in continental margins, while less is known about the net transfer of carbon between estuary and ocean. Less than 0.1 Pg C/yr[-1] is preserved in nearshore marine sediments and a large portion of this is marine-derived (Berner, 1982; Hedges and Keil, 1995). Thus, most of the terrestrial organic carbon must be respired in estuary and coastal waters and sediments (McKee, 2003), or transported somehow to the interior ocean, although few quantitative estimates of either loss term exist. Turbid river plumes are sites of intense remineralization of POC (Aller, 1998). Coastal sediments are sites of elevated carbon diagenesis in a variety of unique ways (benthic photosynthesis; suboxic processes; non-diffusive transport through sediments).

Figure 1.3. Distribution of chlorophyll content of world oceans, based on SeaWIFS climatological data. When presented with linear color-scaling, the dominance of coastal biomass is clear.

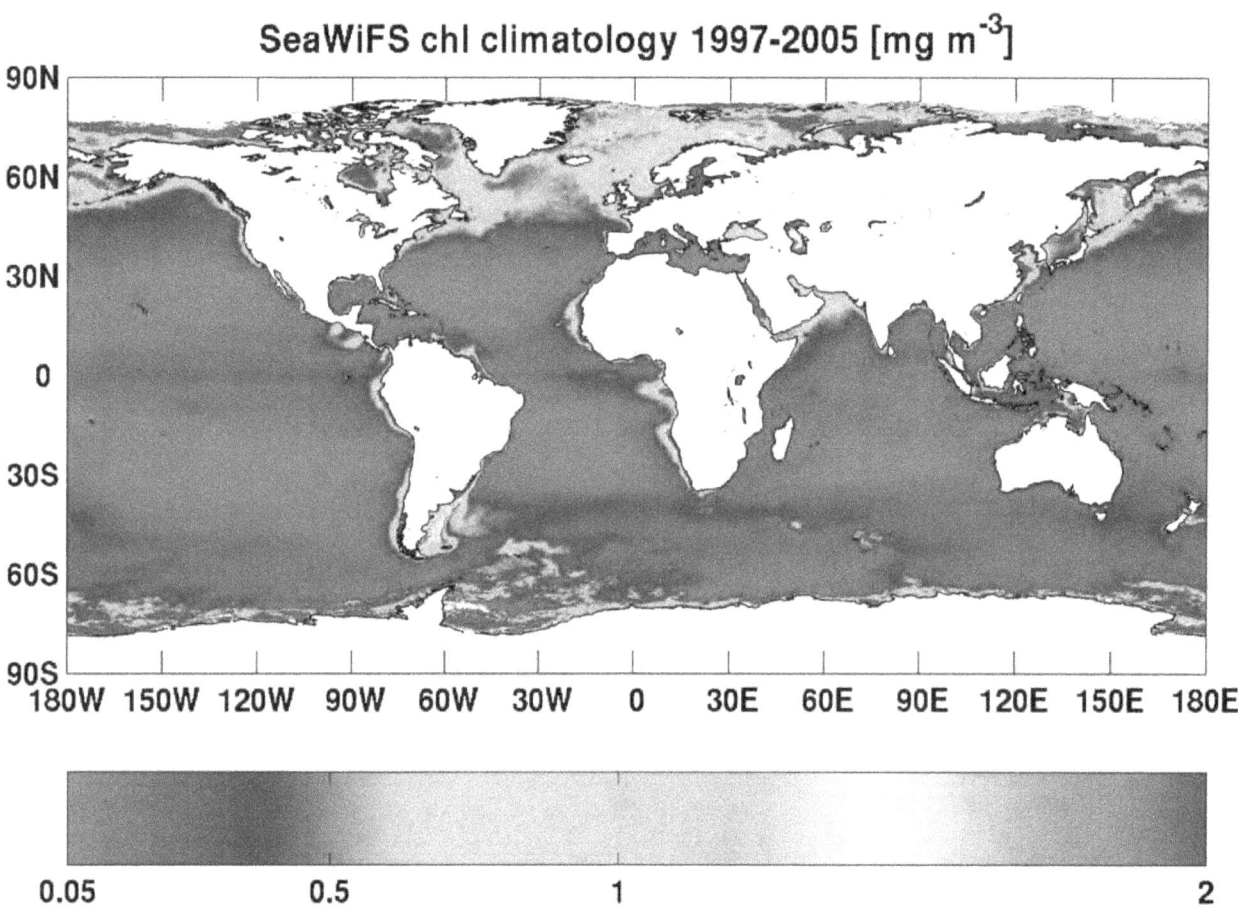

Recent studies suggested that groundwater transport represent an important part of overall terrestrial input to the ocean in some systems (Johannes, 1980; Valiela and Teal, 1979; Valiela et al. 1992; Moore, 1996). This is not totally surprising given that a majority of the world's fresh water is stored underground (Church, 1996). For example, groundwater discharge to lakes and rivers has been used to explain the observed high partial pressure of CO_2 in river waters (Kempe, et al. 1991; Mook and Tan, 1991; Choi et al. 1998). Discharge of groundwater directly into coastal oceans has also been observed (Simmons, 1992), and inputs of bioreactive materials from these systems may exceed river inputs in some coastal regions (Millham and Howes, 1994; Valiela et al. 1978). Cai et al. (2003b) reported that, in the North Inlet (NC, USA) aquifer, DIC and TAlk were several times higher than the river-seawater theoretical dilution line and partial pressure of CO_2 in the groundwater was extremely high (0.05-0.12 atm). Thus even if the overall groundwater discharge rate is only a few percentage of river discharge, the groundwater carbon flux and its influence on coastal sea-to-air CO_2 flux cannot be ignored.

Several other publications have suggested that there is net carbon export from coastal to deep ocean, either by deep water formation, sediment transport, or simply enhanced export productivity (Walsh et al. 1991; Tsunogai et al. 1999; Yool and Fasham, 2001; Muller-Karger et al. 2005; Thomas et al. 2004; Hales et al. 2005, 2006). Inorganic carbon flux from marginal seas to the open ocean may be much higher than river fluxes due to enhanced respiration and or input from highly productive nearshore systems such as marshes (Tsunogai et al. 1999; Cai et al. 2003a; Thomas et al. 2004). These mechanisms are complicated, often requiring extensive transfer between inorganic and organic C pools, and decoupling of particle and water-mass transport. Direct quantification of such sequestration is extremely challenging.

Anthropogenic Impacts on Continental Margins

Given their proximity to the majority of the world's population, continental margins are directly impacted by numerous anthropogenic activities. Not surprisingly these productive regions near shore account for nearly 90% of the world's fish catch (Riley,

1946; Ryther, 1969; Harris, 1980; Steele, 1974; Holligan and Reiners, 1992). Evidence is mounting that human activity is altering nutrient patterns and food web structure (Hallegraeff, 1993; Jorgensen and Richardson, 1996) and these human-induced changes are likely to increase in the coming decades with the projected development along the world's coastlines. For example, more than half of the population of the United States now lives within an hour's drive of the ocean, and current projections expect this to increase in the next decade. This intense human activity is altering global and mesoscale processes through fishing, eutrophication, commercial development, recreation, introduction of exotic species, dredging, the greenhouse effect, etc. (Sherman and Alexander, 1986; Carlton, 1987; Hallegraeff et al. 1991; Pauly and Christensen, 1995; Cloern, 2001; Moncheva et al. 2001). How these changes will impact the overall productivity, and carbon and nitrogen cycles remains an open question for the research community.

Continental margins are also sensitive to the effects of large-scale climate change in a variety of ways aside from simple warming. Freshwater inputs to coastal oceans are predicted to be dramatically impacted, with increases in river discharge at higher latitudes and decreases at lower latitudes (e.g., Manabe et al. 2004). It has been suggested that increased land-ocean temperature gradients in a warming climate will lead to stronger coastal wind forcing (Bakun, 1990). These changes will have obvious impacts on wind- and buoyancy-driven circulation, as well as the supply of terrestrial- and pelagic-source nutrients, that will directly affect carbon cycling processes on the continental margins.

North American Continental Margins

North America's continental margins exemplify the global margins. Stretching nearly from the equator to the north pole, these margins encompass wide ranges in productivity, influences of exchange with the terrestrial and pelagic ecosystems, interactions with the seafloor, atmospheric forcing conditions, and anthropogenic alteration. Observations of surface pCO_2 in these settings make it clear that these environments are significant factors in carbon budgets in the regional and seasonal sense at the very least.

Table 1.1 Annual air-sea CO_2 flux measurement programs in North American continental shelves (positive value is release to the air; negative value is uptake by the sea). Most fluxes are based on direct pCO_2 measurements. Fluxes were calculated using the Wanninkhof (1992) formula. Note that the small annual fluxes in the Arctic shelves are caused by ice coverage while their fluxes in the ice-free season are extremely large (compiled by W.-J. Cai).

Shelf name	Latitude (°N)	CO_2 flux gC m^{-2} yr^{-1}	Researcher	Note
Pacific				
Bering Sea	58	-30	Codispoti & Friederich/ Chen	No annual data; Rough estimate
Vancouver Island	49	-6	Ianson/Wong	Modeling extrapolation
Oregon coast	44.5	-24	Hales/Takahashi/ Van Geen	Upwelling season
Central California	36.7	-6	Friederich/Chavez	El Niño & weak upwelling year
Central California	36.7	18	Friederich/Chavez	La Niña & strong upwelling
Atlantic				
Gulf of Maine	42.8	-29	Salisbury/Vandemark	Limited area
Mid Atlantic Bight (MAB)*	38.5	-20*	DeGrandpre/Takahashi	Whole shelf
South Atlantic Bight (SAB)**	31	24	Cai	Limited area
Gulf coast and further south				
West FL shelf	25	20	Millero	
Mississippi River Plume area	28	-6	Cai/Lohrenz	Not annually integrated
Caribbean Sea	18	-0	Wanninkhof/Olsen	Including deep water area
CARIACO time series	10.5	34	Muller-Karger/Astor	5-year data;$^+$ upwelling site
Arctic	>65	-10	Anderson et al./Murata/ Bates	based on DIC budget
Bering Sea	58	-30	Codispoti & Friederich/ Chen	No annual data; Rough estimate

* Adjusted atmospheric pCO_2.
**Limited to central bight and nearshore area. Recent whole shelf survey suggested a smaller flux.
$^+$fCO$_2$ calculated from pH and alkalinity.

How these margins affect net annual air-sea CO_2 fluxes is less clear. In North America ocean margins, there have been several studies focused on air-sea CO_2 exchange (Table 1.1). In addition, a recent re-examination of the coastal measurements included in Takahashi's global CO_2 database (Chavez et al. 2007; Figure 1.4a), including many new coastal measurements, has been completed. The coastal data includes nearly 10^6 measurements made in North American continental margins since 1979. As such, carbon cycling in the North American continental margins may be better studied than any other margin, with the possible exception of the European Atlantic coasts.

Despite this, spatial and temporal coverage is still quite limited, leaving huge uncertainty as to the net air-sea exchange of CO_2 with the atmosphere. Chavez et al. (2007) binned their data into 1° latitude by 1° longitude by one calendar-month intervals, interpolating where necessary, to create a single, monthly resolved, 'composite-year' of data. In doing this, they found that the net air-sea CO_2 flux from North American coastal waters was about 2 ± 20 Tg C yr^{-1}, but this small net flux was dominated by a large source from low-latitude waters and a nearly compensating large sink in high-latitude waters. As

seen in Figure 1.4b, there are actual measurements in a very low number of calendar months in the pixels in these regions, making the large regional fluxes that determine the total based mostly on interpolation of a very sparse data record. In addition to these undersampled hotspots, the coastal data-set as a whole is fairly sparse. Less than 10% of the pixels have measurements in 12 calendar months, and the average number of calendar months in a given pixel during which measurements were made is about 2.5. Clearly, there is a demand for more surface CO_2 measurement in the waters of the North American continental margin.

Questions regarding other aspects of the carbon cycling in North American margins are less well-constrained. The issues regarding the discharge of riverine carbon to estuaries relative to the delivery of this material to the coastal oceans are significant. There have been no direct measurements of transfer of carbon between coastal and pelagic oceans. Rates of water column and sediment respiration of both marine and terrestrially derived organic matter are mostly unknown. The importance of submarine groundwater discharge to the coastal margin carbon cycle is a subject of mere speculation.

Figure 1.4. a) Locations of coastal CO_2 measurements in the LDEO global CO_2 database, as compiled by Chavez et al.. (2007). b) Locations of the 1° x 1° pixels within 300 km of the coastline used to create the calendar-year composite. Colors indicate the number of calendar months in each pixel in which actual measurements were made.

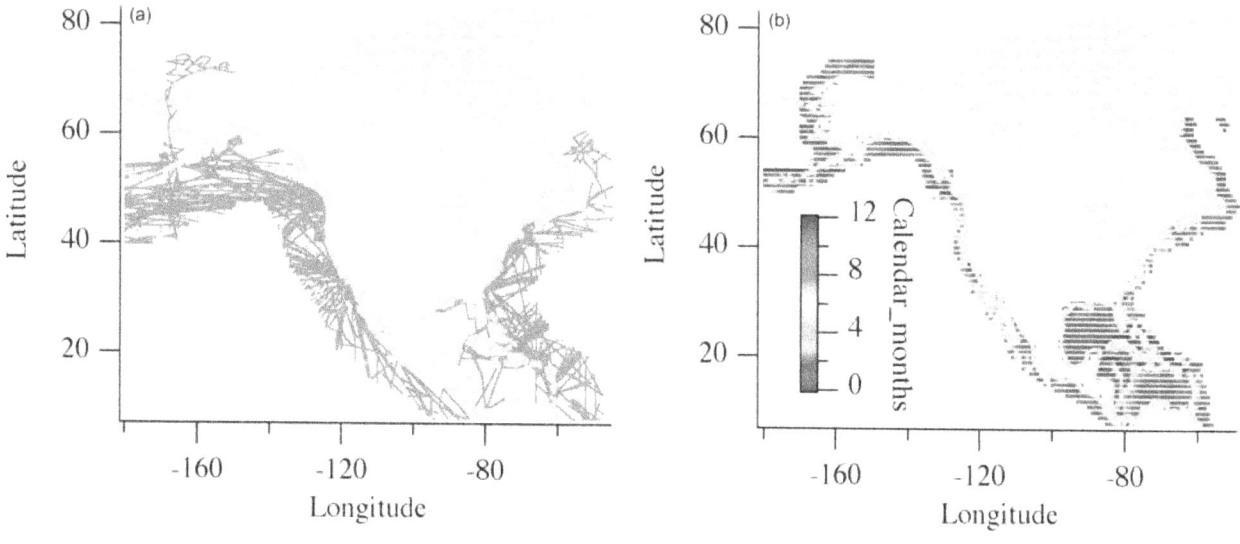

The above discussion suggests that there are several key unresolved questions regarding the carbon cycle in North America's continental margins. These are:

- **Are coastal oceans a net source or sink for atmospheric CO_2?**

It is safe to say that on a regional and seasonal basis, the North American continental margins generate significant air-sea CO_2 fluxes. Large dynamic ranges, including large fluxes of opposite sign, short temporal and spatial scales of variability, and undersampling in key source/sink regions prevent estimation of either the sign or magnitude of the net annual flux. The mechanisms that drive these fluxes, and the factors that control their spatial and temporal patterns are poorly understood, leaving little ability to predict how these will change in the face of anthropogenic perturbations and climate change.

- **What is the net transfer of carbon between coastal and open oceans?**

Several studies have suggested transfer of carbon from the coastal oceans to the deep ocean interior, via dissolved and particulate transport pathways, in organic and inorganic carbon phases. There have been no direct measurements of such transport.

- **What is the net source and form of carbon and nutrients from rivers through estuaries to the ocean?**

While there are significant uncertainties in the magnitudes and modes of riverine carbon discharge fluxes, it is at least known that rivers deliver significant quantities of carbon and nutrients to estuaries. Much less is known about the actual delivery of this material to the coastal oceans at the ocean estuary boundary. Estuaries are sites of intense transformation, and direct determination of fluxes through tidally influenced estuary mouths is extremely challenging.

- **What are the net exchanges of carbon through the coastal seafloor?**

Accumulation of carbon in margin sediments is a globally significant pathway for long-term carbon sequestration, yet this sink represents a small portion of the autochthonous production in and allochtonous delivery to coastal margins. Benthic remineralization appears to be a small part of water column sources. The relatively recent suggestions of elevated benthic photosynthesis and submarine groundwater discharge only complicate our understanding of the transfer of carbon through this interface.

- **What is the net processing of terrestrial and marine carbon in the coastal ocean?**

Closure of macroscopic carbon budgets in margin settings is difficult, with the imbalances generally ascribed to some unobservable process. An important factor in this uncertainty is our lack of understanding of net carbon diagenesis in the water column and sediments of the coastal ocean. Primary productivity estimates are of limited applicability in regions with widely varying f-ratio and close coupling with the sea floor, let alone those strongly impacted by delivery of terrestrial carbon. An understanding of net community diagenesis is critical.

These questions can be summarized as a lack of knowledge of the net fluxes of total carbon through key interfaces defining the ocean margins, and ignorance of the net diagenesis of total carbon in the region bounded by these interfaces. In recognition of the importance of carbon cycling in the North American continental margins, we staged a workshop to address the state of continental margin carbon cycle science. This workshop is described in the following chapters.

References

Ahrens, C.D., 2007. *Meteorology Today: An Introduction to Weather, Climate, and the Environment.* 8th edition. Brooks/Cole, Belmont, CA, 537 pp.

Aller, R.C., 1998. Mobile deltaic and continental shelf muds as suboxic, fluidized bed reactors. *Marine Chemistry*, 61:143-155.

Bakun, A., 1990. Global climate change and intensification of coastal ocean upwelling. *Science*, 247:198-201.

Baker, D.F., et al. 2006: TransCom 3 inversion intercomparison: Impact of transport model errors on the interannual variability of regional CO2 fluxes, 1988–2003. *Global Biogeochemical Cycles*, 20:GB1002, doi:10.1029/2004GB002439.

Bates, N.R., 2006. Air-sea CO_2 fluxes and the continental shelf pump of carbon in the Chukchi Sea adjacent to the Arctic Ocean. *Journal of Geophysical Research*, 111:C10013, doi:10.129/2005JC003083.

Berner, R.A., 1982. Burial of organic carbon and pyrite sulfur in the modern ocean: Its geochemical and environmental significance. *American Journal of Science*, 282:451-473.

Bianchi, A., L. Bianucci, A. Piola, D. Ruiz-Pino, I. Schloss, A. Poisson, and C. Balestrini, 2005. Vertical stratification and air-sea CO_2 fluxes in the Patagonian shelf. *Journal of Geophysical Research*, 110:C07003, doi:10.1029/2004JC002488.

Borges, A.V., 2005. Do we have enough pieces of the jigsaw to integrate CO_2 fluxes in the coastal ocean? *Estuaries*, 28:3-27.

Borges, A.V., B. Delille, and M. Frankignoulle, 2005. Budgeting sinks and sources of CO_2 in the coastal ocean: Diversity of ecosystems counts. *Geophysical Research Letters*, 32:L14601, doi:10.1029/2005GL023053.

Borges, A.V., L.-S. Schiettecatte, G. Abril, B. Delille, and F. Gazeau, 2006. Carbon dioxide in European coastal waters. *Estuarine, Coastal and Shelf Science*, 70:375-387.

Cai, W.-J., Z. Wang, and Y. Wang, 2003a. The role of marsh-dominated heterotrophic continental margins in transport of CO_2 between the atmosphere, the land-sea interface and the ocean. *Geophysical Research Letters*, 30:1849.

Cai, W.-J., Y. Wang, J. Krest, and W.S. Moore, 2003b. The geochemistry of dissolved inorganic carbon in a surficial groundwater aquifer in North Inlet, South Carolina, and the carbon fluxes to the coastal ocean. *Geochimica et Cosmochimica Acta*, 67:631-637, doi:10.1016/S0016-7037(02)01167-5.

Cai, W.-J., M. Dai, and Y. Wang, 2006. Air-sea exchange of carbon dioxide in ocean margins: A province based synthesis. *Geophysical Research Letters*, 33:L12603, doi:10.1029/2006GL026219.

Carlton, J.T., 1987. Patterns of transoceanic marine biological invasions in the Pacific Ocean. *Bulletin of Marine Science*, 41:452-465

Chavez, F.P., and J. R. Toggweiler, 1995. Physical estimates of global new production: The upwelling contribution. In: *Upwelling in the Ocean: Modern Processes and Ancient Records.* [C.P. Summerhayes, K.-C. Emeis, M.V. Angel, R.L. Smith, and B. Zeitschel (eds.)]. Wiley, pp. 313-320.

Chavez, F.P., C.A. Collins, A. Huyer, and D.L. Mackas, 2002. El Niño along the west coast of North America. *Progress in Oceanography*, 54:1-5.

Chavez, F.P., T. Takahashi, W.-J. Cai, G. Friederich, B. Hales, R. Wanninkhof, and R. Feely, 2007. Coastal oceans. In: *The First State of the Carbon Cycle Report (SOCCR): The North American Carbon Budget and Implications for the Global Carbon Cycle.* [A.W. King, L. Dilling, G.P. Zimmerman, D.M. Fairman, R.A. Houghton, G. Marland, A.Z. Rose, and T.J. Wilbanks (eds.)]. A report by the U.S. Climate Change Science Program and the Subcommittee on Global Change Research, Washington, DC, pp. 157-166. Available at http://www.climatescience.gov/Library/sap/sap2-2/final-report/default.htm.

Chen, C.-T.A., A. Andreev, K.-R. Kim, and M. Yamamoto, 2004. Roles of continental shelves and marginal seas in the biogeochemical cycles of the North Pacific Ocean. *Journal of Oceanography*, 60:17-44.

Choi, J., S.M. Hulseapple, M. H. Conklin and J. W. Harvey, 1998. Modeling CO_2 degassing and pH in a stream–aquifer system. *Journal of Hydrology*, 209:297-310.

Church, T.M., 1996. An underground route for the water cycle. *Nature*, 380:579-580.

Cloern, J.E., 2001. Our evolving conceptual model of the coastal eutrophication problem. *Marine Ecology Progress Series*, 210:223-253.

DeGrandpre, M.D., T.R. Hammar, G.J. Olbu, and C.M. Beatty, 2002. Air-sea CO_2 fluxes on the US Middle Atlantic Bight. *Deep-Sea Research II*, 49:4355-4367.

Ducklow, H.W., and S.L. McAllister, 2005. The biogeochemistry of carbon dioxide in the coastal oceans. In: The Global Coastal Ocean: Multiscale Interdisciplinary Processes [K.H. Brink and A.R. Robinson (eds.)]. The Sea, Volume 13. Harvard University Press, Cambridge, MA.

Frankignoulle, M., and A.V. Borges, 2001. European continental shelf as a significant sink for atmospheric carbon dioxide. *Global Biogeochemical Cycles*, 15:569-576.

Friederich, G.E., P.M. Walz, M.G. Burczynski, and F.P. Chavez, 2002. Inorganic carbon in the central California upwelling system during the 1997–1999 El Niño–La Niña event. *Progress in Oceanography*, 54:185-203.

Hales, B., T. Takahashi, and L. Bandstra, 2005. Atmospheric CO_2 uptake by a coastal upwelling system, *Global Biogeochemical Cycles*, 19, doi:10.1029/2004GB002295.

Hales, B., L. Karp-Boss, A. Perlin, and P. Wheeler, 2006. Oxygen production and carbon sequestration in an upwelling coastal margin. *Global Biogeochemical Cycles*, 20:GB3001, doi:10.1029/2005GB002517.

Hallegraeff, G., 1993. A review of harmful algal blooms and their apparent global increase. *Phycologia*, 32:79-99.

Hallegraeff, G.M., C.J. Bolch, S.I. Blackburn, and Y. Oshima, 1991. Species of the toxigenic dinoflagellate genus Alexandrium in southeastern Australian waters. *Botanica Marina*, 34:575-587.

Harris, G.P., 1980. Temporal and spatial scales in phytoplankton ecology: Mechanisms, methods, models, and management. *Canadian Journal of Fisheries and Aquatic Sciences*, 37:877-900.

Hedges, J.I., and R.G. Keil, 1995. Sedimentary organic matter preservation: an assessment and speculative synthesis. *Marine Chemistry*, 49:81-115.

Holligan, P.M., and W.A. Reiners, 1992. Predicting the responses of the coastal zone to global change. *Advances in Ecological Research*, 22:211-255.

Ianson, D., and S.E. Allen, 2002. A two-dimensional nitrogen and carbon flux model in a coastal upwelling region. *Global Biogeochemical Cycles*, 16:1011, doi:10.1029/2001GB001451.

Ianson, D., S.E. Allen, S.L. Harris, K.J. Orians, D.E. Varela, and C.S. Wong, 2003. The inorganic carbon system in the coastal upwelling region west of Vancouver Island, Canada. *Deep-Sea Research I*, 50:1023-1042.

Johannes, R.E., 1980. The ecological significance of the submarine discharge of groundwater. *Marine Ecology Progress Series*, 3:365-373.

Jorgensen, B.B., and K. Richardson, 1996. *Eutrophication in Coastal Marine Ecosystems*. American Geophysical Union Coastal and Estuarine Studies 52, Washington, DC, 273 pp.

Kempe, S., M. Pettine, and G. Cauwet, 1991. Biogeochemistry of Europe rivers. In: *Biogeochemistry of Major World Rivers* [E.T. Degens, S. Kempe, and J.E. Richey (eds.)]. SCOPE 42. John Wiley, New York, pp. 169-211.

Lohrenz, S.E., and W.-J. Cai, 2006. Satellite ocean color assessment of air-sea fluxes of CO_2 in a river-dominated coastal margin. *Geophysical Research Letters*, 33:L01601, doi:10.1029/2005GL023942.

Lueker, T.J., S.J. Walker, M.K. Vollmer, R.F. Keeling, C.D. Nevison, R.F. Weiss, and H.E. Garcia, 2003. Coastal upwelling air-sea fluxes revealed in atmospheric observations of O_2/N_2, CO_2 and N_2O. *Geophysical Research Letters*, 30:1292, doi:10.1029/2002GL016615.

Manabe, S., P.C.D. Milly, and R. Wetherald, 2004. Simulated long-term changes in river discharge and soil moisture due to global warming. *Hydrological Sciences Journal*, 49:625-642.

McKee, B., 2003. *RiOMar: The Transport, Transformation and Fate of Carbon in River-Dominated Ocean Margins*. Report of the RiOMar Workshop, 1-3 November 2001. Tulane University, New Orleans, LA.

Millham N.P., and B.L. Howes, 1994. Freshwater flow into a coastal embayment: Groundwater and surface water inputs. *Limnology and Oceanography*, 39:1928-1944.

Moncheva, S., O. Gotsis-Skretas, K. Pagou, and A. Krastev, 2001. Phytoplankton blooms in Black Sea and Mediterranean coastal ecosystems subjected to anthropogenic eutrophication: Similarities and differences. *Estuarine, Coastal and Shelf Science*, 53:281-295, doi:10.1006/ecss.2001.0767.

Mook, W.G., and F.C. Tan, 1991. Stable carbon isotopes in rivers and estuaries. In: *Biogeochemistry of Major World Rivers* [E.T. Degens, S. Kempe, and J. E. Richey (eds.)]. SCOPE 42. John Wiley, New York, pp. 245-264.

Moore W.S., 1996. Large groundwater inputs to coastal waters revealed by 226Ra enrichments. *Nature*, 380:612-614.

Moran, S.B., M.A. Charette, S.M. Pike, and C.A. Wicklund, 1999. Differences in seawater particulate organic carbon concentration in samples collected using small-volume and large-volume methods: the importance of DOC adsorption to the filter blank. *Marine Chemistry*, 67:33-42.

Muller-Karger, F.E., R. Varela, R. Thunell, R. Luerssen, C. Hu, and J.J. Walsh, 2005. The importance of continental margins in the global carbon cycle. *Geophysical Research Letters*, 32:L01602, doi:10.1029/2004GL021346.

Pauly, D., and V. Christensen, 1995. Primary productivity required to support fisheries. *Nature*, 374:255-257.

Peterson, W.T., and F. B. Schwing, 2003. A new climate regime in northeast Pacific ecosystems. *Geophysical Research Letters*, 30:1896, doi:10.1029/2003GL017528.

Riley, G.A., 1946. Factors controlling phytoplankton populations on Georges Bank. *Journal of Marine Research*, 6:54-73.

Ryther, J.H., 1969. Photosynthesis and fish production in the sea. *Science*, 166:72-76.

Sherman, K., and L.M. Alexander, 1986. *Variability and Management of Large Marine Ecosystems*. Westview Press, Boulder, CO.

Simmons, G.M., 1992. Importance of submarine groundwater discharge (SGWD) and seawater cycling to material flux across sediment/water interfaces in marine environments. *Marine Ecology Progress Series*, 84:173-184.

Smith, S.V., and J.T. Hollibaugh, 1993. Coastal metabolism and the oceanic organic carbon balance. *Reviews of Geophysics*, 31:75-89

Steele, J.H., 1974. *The Structure of Marine Ecosystems*. Harvard University Press, Cambridge, MA.

Takahashi, T., S.C. Sutherland, C. Sweeney, A. Poisson, N. Metzl, B. Tilbrook, N. Bates, R. Wanninkhof, R.A. Feely, C. Sabine, J. Olafsson, and Y. Nojiri, 2002. Global sea-air CO_2 flux based on climatological surface ocean pCO_2, and seasonal biological and temperature effects. *Deep Sea Research II*, 49:1601-1622.

Thomas, H., Y. Bozec, K. Elkalay, and H.J.W. de Baar, 2004. Enhanced open ocean storage of CO_2 from shelf sea pumping. *Science*, 304:1005-1008.

Tsunogai, S., S. Watanabe, and T. Sato, 1999. Is there a "continental shelf pump" for the absorption of atmospheric CO_2? *Tellus B*, 51:701-712.

Valiela, I., and J.M. Teal, 1979. The nitrogen budget of a salt marsh ecosystem. *Nature*, 280:652-656.

Valiela, I., J.M. Teal, S. Volkman, D. Shafer, and E.J. Carpenter, 1978. Nutrient and particulate fluxes in a salt marsh ecosystem: Tidal exchanges and inputs by precipitation and groundwater. *Limnology and Oceanography*, 23:798-812.

Valiela, I., K. Foreman, M. LaMontagne, D. Hersh, J. Costa, P. Peckol, B. DeMeo Anderson, C. D'Avanzo, M. Babione, C.-H. Sham, J. Brawley, and K. Lajtha, 1992. Coupling of watersheds and coastal waters: Sources and consequences of nutrient enrichment in Waquoit Bay, Massachusetts. *Estuaries*, 15:443-457.

Vandemark, D., J. Salisbury, C. Hunt, W. McGillis, J. Campbell, and F. Chai, 2006. The coastal Gulf of Maine as an atmospheric CO_2 sink with strong seasonal riverine control. *Eos*, 87(36), Ocean Sciences Meeting Supplement, Abstract OS35G-15.

Walsh, J.J., D.A. Dieterle, and J.R. Pribble, 1991. Organic debris on the continental margins: a simulation analysis of source and fate, *Deep-Sea Research*, 38:805-828.

Wanninkhof, R., 1992. Relationship between wind-speed and gas exchange over the ocean. *Journal of Geophysical Research*, 97:7373-7382.

Yool, A., and M.J.R. Fasham, 2001. An examination of the "continental shelf pump" in an open ocean general circulation model. *Global Biogeochemical Cycles*, 15:831-844.

Workshop Motivation and Description

Burke Hales
College of Oceanic and Atmospheric Sciences
Oregon State University

Wei-Jun Cai
University of Georgia

B. Greg Mitchell
Scripps Institution of Oceanography
University of California, San Diego

Chris Sabine
Pacific Marine Environmental Laboratory
National Oceanographic and Atmospheric Administration

Oscar Schofield
Institute of Marine and Coastal Sciences
Rutgers University

Background

Two national carbon research programs, the North American Carbon Program (NACP) and Ocean Carbon and Climate Change (OCCC) (Wofsy and Harriss, 2002; Doney et al. 2004; Denning et al. 2005) have recognized the importance of carbon-cycle study in the ocean margins. The NACP is a national plan for carbon cycle research focused on measuring and understanding sources and sinks of CO_2, CH_4, and CO in North America and adjacent oceans. The NACP plan addresses four fundamental topics:

1. What is the carbon balance of North America and adjacent oceans? What are the geographic patterns of CO_2, CH_4, and CO fluxes? How is the balance changing over time? ("Diagnosis")
2. What processes control the sources and sinks of CO_2, CH_4, and CO, and how do the controls change with time? ("Attribution/Process")
3. Are there potential surprises (could sources increase or sinks disappear)? ("Prediction")
4. How can we enhance and manage long-lived carbon sinks ("Sequestration"), and provide resources to support decision makers? ("Decision support")

Ocean Carbon and Climate Change (OCCC) is an implementation strategy for US ocean carbon research, and has the broader goals of determining (1) the global inventory, geographic distribution, and temporal evolution of anthropogenic CO_2 in the oceans; (2)

the magnitude, spatial pattern, and variability of air-sea CO_2 flux; (3) the major physical, chemical, and biological feedback mechanisms and climate sensitivities for ocean organic and inorganic carbon storage; and (4) the scientific basis for ocean carbon mitigation strategies.

There are obviously significant overlaps between the NACP and OCCC. These were recognized and the respective plans were designed to complement each other to provide a seamless integration of ocean, atmosphere, and terrestrial carbon cycle research in the United States and adjacent ocean basins. One major region of overlap is in the coastal oceans. Both programs have coastal carbon studies as a major element of their plans. The proposed division of responsibilities gives primary responsibility for land-ocean interactions to NACP, primary responsibility for open ocean-coastal ocean interactions to OCCC, and both programs would collaborate to understand the continental shelf regions. Although a few studies like the NSF CoOP (Coastal Ocean Processes) and initiatives like the RiOMar (River-dominated Ocean Margins) programs have synthesized, or will synthesize, results from the North American coast, what is needed for NACP and OCCC is a **coordinated large-scale coastal carbon research effort.** Both programs note that: "*The first step for developing a specific coastal ocean plan is to organize a cross-disciplinary NACP/OCCC workshop including coastal oceanographers currently working in the North American continental margins to outline the existing programs and opportunities for collaboration and to refine the needs of the NACP and OCCC and develop a detailed strategy for each region.*"

In the following, we describe the development, execution, and results from a workshop intended to synthesize our understanding of the spatial and temporal scales of variability on the North American continental margins with the purpose of determining the observational and modeling requirements for meeting the research goals of the NACP and OCCC. The goal of this workshop was to bring the coastal community together to discuss the scales of variability that are relevant for addressing continental-scale aspects of carbon cycling, to look for synergies among the on-going and planned coastal programs, and to develop a realistic plan for implementing a coastal carbon program.

Workshop Genesis

In the fall of 2004, Burke Hales, Wei-Jun Cai, Greg Mitchell, Chris Sabine, and Oscar Schofield together submitted a proposal to the Carbon Cycle Interagency Working Group to organize and execute the North American Continental Margins Workshop. Funds were awarded jointly by NASA, NOAA, and NSF through UCAR in spring 2005 to support a three-day workshop with approximately 50 participants.

Early in the process, the organizers recognized that the major geographical subregions of North America's continental margins—the Pacific, Atlantic, Arctic, and Gulf of Mexico coasts, as well as the Laurentian Great Lakes and river/estuarine systems—were each distinguished by the predominance of processes that drive carbon cycling, and these distinctions shaped the workshop from the outset. These are each discussed in detail in the regional chapters that follow, but can be briefly summarized here:

The Atlantic Coast consists of a broad, shallow shelf separated from the low-nutrient open North Atlantic by a strong western boundary current and bounded on the landward side by extensive estuarine systems—salt marshes in the south, large embayments at mid-latitudes, and riverine systems to the north. Upwelling is limited, leading to lower overall water-column productivity and a greater role of benthic processes in the net carbon budget than in some other regions. Strong episodic storm events play a significant role in shaping water-column productivity and sediment transport.

The Pacific Coast, in contrast, consists of a narrower shelf delineated from the high-nutrient open North Pacific by weaker eastern boundary currents, and on the landward side by exposed open coastlines with limited river input in the south and by systems of islands and embayments with large freshwater input in the north. Wind-driven circulation is critical in shaping carbon cycling in this region, with strong upwelling at mid and low latitudes generating large net production. Mesoscale processes such as eddies and short-term wind fluctuations are significant factors driving carbon export to the open ocean.

The Gulf Coast is a large enclosed tropical sea with broad shelves that is strongly influenced by river input along its northern and southern oceans by the largest rivers in the United States (Mississippi) and Mexico (Usumucinta). This system is strongly influenced additionally by episodic strong storm events.

The Arctic Coast (including the Bering Sea) contains the greatest fraction of shelf area in North America. It is unique relative to other marginal regions in the predominance of sea-ice processes, and as the only region where deepwater formation is significant. The Arctic may be the most sensitive of North America's margins to long-term climate change, as the polar ice caps melt, permafrost thaws, and Arctic river discharge increases.

The Laurentian Great Lakes, North America's inland ocean, are the only purely freshwater environment included in this review of continental margins, and carbon transport between the Great Lakes and the open ocean is unidirectionally to the ocean through the Saint Lawrence River/estuary system. Relative to other marginal systems, degradation of terrestrial DOC is a more significant part of the lakes' carbon cycle. This system is strongly impacted by anthropogenic activities that have altered the natural carbon cycle, including pollution-driven eutrophication and introduction of invasive species.

River/Estuary Systems are the sites of active transfer of dissolved and particulate carbon from terrestrial to oceanic reservoirs. Despite some uncertainty, it is accepted that rivers discharge a large amount of carbon to their estuaries; however, the state of river/estuary research leaves the flux of carbon between estuaries and oceans largely unconstrained. Estuaries are areas of complex carbon cycling, with intimate coupling between sediment transport, respiration, and productivity driving the diagenesis of terrestrial- and marine-source carbon alike.

The co-organizers recognized that the expertise required to synthesize carbon cycling in these diverse settings required external input, and selected a panel of external expert advisors with established research records in the major regions of the coastal oceans of North America. These advisors were: Rick Jahnke, Atlantic Coast; Francisco Chavez, Pacific Coast; Steve Lohrenz, Gulf Coast; Jackie Grebmeier, Arctic Coast; Jim Bauer, Rivers/Estuaries; and Val Klump, Great Lakes. They assisted the co-organizers with selection of invitees and development of the agenda. Invitations were sent to over 50 researchers and agency representives from around the world representing a variety of expertise.

Workshop Objectives

The objectives of the workshop were as follows:

- Summarize and synthesize the 'state of the art' regarding C cycling on the continental margins (the knowns).
- Identify the key processes that shape regional C cycling.
- Identify the most pressing uncertainties in our ability to estimate coastal C fluxes (the known unknowns).
- Hypothesize about potential responses of coastal systems (and inherent C cycling) to global change.
- Offer guidelines for future coastal research programs.
- Present these in a formal report to US funding agencies and the IWG.

Workshop Description

The North American Continental Margins workshop was held at the Boulder Millenium Hotel in Boulder, CO from 21-23 September, 2005. Although hurricanes Katrina and Rita significantly depleted participation by Gulf Coast experts, it was attended by 46 participants, who represented six nations, 37 scientific institutions, and two US funding agencies (Table 2.1).

Meeting Agenda

The workshop opened with a plenary session designed to provide the attendees background on the need for a continental margins synthesis and planning effort, including summaries of the OCCC and NACP programs, and the potential overlap of the two programs in the continental margins. Following that was a plenary session in which the regional experts presented summaries of the carbon cycle in each region.

Following these sessions, a series of pre-planned breakout groups was convened, covering the six margin regions, in addition to three sessions covering in-water measurement strategies, remote sensing applications, and modeling and synthesis efforts for study of the carbon cycle in marginal settings. Each group began with several brief, informal presentations offered by the meeting participants. These were intended to be presentations of ongoing research in each subject area that might not have been covered in the plenary sessions. Following these presentations, the groups discussed and summarized the state-of-the-art in each session topic, shortcomings, and recommendations for future research. Following each breakout session were plenary reports on those discussions by the group moderators.

One additional breakout session on the last day of the workshop consisted of three groups that were defined by the meeting participants, covering topics that had not been adequately addressed in the first two days. These included a discussion of boundaries of the coastal margins and division of the margins into a finite set of process-defined subregions; a discussion of synthesis of existing datasets through the prism of carbon cycling; and a discussion of approaches for integrated modeling and process study.

The meeting closed with a discussion of potential agency approaches to supporting research in margins. A detailed agenda follows on page 18.

Table 2.1. Workshop participants.

Participant	Affiliation	Participant	Affiliation
Bob Aller	SUNY-Stonybrook	Ruben Lara	CICESE-Mexico
Nick Bates	Bermuda Biological Station for Research	Steve Lentz	Woods Hole Oceanographic Institution
Jim Bauer[2]	Virginia Institute of Marine Science	Fred Lipschulz[3]	National Science Foundation
Ron Benner	University of South Carolina	KK Liu	National Central University, Taiwan
Paula Bontempi[3]	NASA, Headquarters	Steve Lohrenz[2]	University of Southern Mississippi
Alberto Borges	Universite de Liege, Belgium	Amala Mahadevan	Boston University
Wei-Jun Cai[1]	University of Georgia	Wade McGillis	Lamont-Doherty Earth Observatory
Francisco Chavez[2]	Monterey Bay Aquarium and Research Institute	Brent McKee[4]	Tulane University
C.T.A. Chen	National Sun Yat-Sen University, Taiwan	Galen McKinley	University of Wisconsin at Madison
Lou Codispoti	University of Maryland	Greg Mitchell[1]	Scripps Institution of Oceanography
Jim Cotner	University of Minnesota at St. Paul	Paco Ocampo	CICESE-Mexico
Mike DeGrandpre	University of Montana	Tsung-Hung Peng[4]	NOAA, Atlantic Ocean Marine Laboratory
Carlos DelCastillo[4]	NASA Stennis	Pete Raymond	Yale University
Al Devol	University of Washington	Clare Reimers	Oregon State University
Brian Eadie	NOAA, Great Lakes Environmental Research Laboratory	Don Rice[3,4]	National Science Foundation
Dick Feely	NOAA, Pacific Marine Environmental Laboratory	Chris Sabine[1]	NOAA, Pacific Marine Environmental Laboratory
Gilberto Gaxiola	CICESE-Mexico	Oscar Schofield[1]	Rutgers University
Miguel Goni[4]	Oregon State University	Sybil Seitzinger	Rutgers University
Jackie Grebmeier[2]	UT-Knoxville	Dave Siegel	UC Santa Barbara
Niki Gruber	UCLA	Margaret Squires	University of Waterloo, Canada
Burke Hales[1]	Oregon State University	Ted Strub	Oregon State University
Chuck Hopkinson	Marine Biological Laboratory	Colm Sweeney	National Center for Atmospheric Research
Debbie Ianson	Department of Fisheries and Oceanography-Canada	Kathy Tedesco	NOAA
Rick Jahnke[2]	Skidaway Institute of Oceanography	Peter Thornton	University Center for Atmospheric Research
Ken Johnson	Monterey Bay Aquarium and Research Institute	Rick Wanninkhof	NOAA, Atlantic Ocean Marine Laboratory
Val Klump[2]	University of Wisconsin at Milwaukee	Jim Waples	University of Wisconsin at Milwaukee

[1] Workshop co-organizer.   [3] Agency representative
[2] External advisor   [4] Could not attend due to Hurricane Katrina

Agenda
Day I, 21 September

Session IA: Plenary: OCCC/NACP coastal studies, beginning at 08:00

08:00-08:30   Welcome; presentation of agenda, meeting logistics (Hales)
08:30-09:00   OCCC overview (Feely)
09:00-09:30   NACP overview (Thornton)
09:30-10:00   OCCC/NACP joint coastal studies (Sabine)

Session IB: Plenary: State of art in 'regions' 1-6, beginning at 10:30 on 21 September

10:30-11:00   Plenary presentation: Region 1—East Coast (Rick Jahnke)
11:00-11:30   Plenary presentation: Region 2—West Coast (Francisco Chavez)
11:30-12:00   Plenary presentation: Region 3—Gulf Coast (Steve Lohrenz)
12:00-12:30   Plenary presentation: Region 4—Arctic Shelf (Jackie Grebmeier)
13:30-14:00   Plenary presentation: Region 5—Great Lakes (Val Klump)
14:00-14:30   Plenary presentation: Region 6—North American Rivers (Jim Bauer)

Session IC: Breakout Groups 1: Defining science questions in 'regions' 1-3, beginning at 14:30

Atlantic Coast
moderated by Chuck Hopkinson, Mike DeGrandpre

Brief science presentations:

Mike DeGrandpre: Air-sea fluxes on the Middle Atlantic Bight

Joe Salisbury: Carbon cycling and optics in the Gulf of Maine: Observations and modeling

Rick Jahnke: Unexpected source for Fe and sink for nitrate on continental shelves

Wei-Jun Cai: The terrestrial vs. oceanic inputs in regulating CO_2 fluxes in the South Atlantic Bight

Pacific Coast
moderated by Clare Reimers, Raphe Kudela

Brief science presentations:

Ted Strub: Offshore propagation of eddies along the coast of the Western U.S.

Burke Hales: Quantitative export of coastal net productivity to the deep ocean

Ruben Lara: Carbon sources and sinks in the continental margins of the Mexican Pacific

Gilberto Gaxiola: Carbon process monitoring in temperate and subtropical Mexican coastal waters (IMECOCAL and PROCOMEX)

Niki Gruber: Diurnal cycling and the air-sea exchange of CO_2

Gulf Coast
moderated by Wei-Jun Cai, Rik Wanninkhof

Brief science presentations:

Ron Benner: Terrigeneous DOC and photo-oxidation

Steve Lohrenz: Satellite observations of the northern Gulf of Mexico

Sybil Seitzinger: Implications of N removal by denitrification for the net balance of C in the coastal ocean

Chen-Tung Arthur Chen: Air-sea exchange of carbon in the marginal seas

Bob Aller: Mobile muds as sedimentary C incinerators

End Day I @ 18:00

Day II, 22 September

Session IIA: Plenary: Breakout group summary, beginning at 08:00.

08:00-09:00   Plenary: Brief summary presentations from Breakout Groups 1

Session IIB: Breakout Groups 2: Defining science questions in 'regions' 4-6, starting at 09:00

Arctic Coast
moderated by Al Devol, Ron Benner

Brief science presentations:

Al Devol: Nitrogen cycling in Arctic and Bering sediments

Ron Benner: Phytoplankton production of DOM

Lou Codispoti: The conveyor belt short circuit in the Chukchi Sea

Jackie Grebmeier: The northern Bering Sea: An ecosystem in change

Nick Bates: Carbon cycling (and anthropogenic CO_2) in the Chukchi Sea and Arctic Ocean

Great Lakes
moderated by Jim Waples, Brian Eadie

Brief science presentations:

Margaret Squires: Distribution of pCO_2 in Lake Erie surface waters and implications for carbon limitation of phytoplankton

Jim Cotner: Carbon and photochemical dynamics in Lake Superior

Rivers and Estuaries
moderated by Peter Raymond, Sybil Seitzinger

Brief science presentations:

Alberto Borges: Do we have enough pieces of the jigsaw puzzle to integrate CO_2 fluxes in the coastal ocean?

Pete Raymond: The composition and transport of organic carbon in rainfall: Insights from the natural (^{13}C and ^{14}C) isotopes of carbon

Pete Raymond: The age of carbon being exported by arctic rivers

KK Liu: Deposition of particulate organic carbon in three types of margins: the role of sediment flux

Session IIC: Plenary: Breakout group summary, beginning at 13:30

13:30-14:30   Plenary: Brief summary presentations from Breakout Groups 2

Session IID: Breakout Groups 3: Technical advances—Observation and synthesis at appropriate temporal and spatial scales

14:30-18:00   Breakout Groups 3: Technical advances—Observation and synthesis at appropriate temporal and spatial scales

In-water measurement technology and strategies
moderated by Wade McGillis, Ken Johnson

Brief science presentations:

Wade McGillis: An overview: Recent developments in coastal ocean CO_2 studies on the US East coast continental shelf

Lou Codispoti: Recent advances in autonomous nutrient monitoring and telemetry

Ken Johnson: Primary production and carbon cycling monitored via in situ nutrient sensors

Colm Sweeney: High accuracy autonomous pCO_2 measurements for coastal air-sea fluxes and the NACP

Sybil Seitzinger: Opening the Black Box of DOM - molecular level characterization of DOM sources and utilization

Bob Aller: New planar optodes for high resolution, two-dimensional measurements of solute distributions in sediment and overlying water

Remote sensing approaches
moderated by Dave Siegel, Ted Strub

Brief science presentations:

Ted Strub: The Cooperative Institute for Oceanographic Satellite Studies (CIOSS)

Greg Mitchell: Spatially uncoupled response to El Niño of satellite-derived chlorophyll, productivity and export flux along the west coast of North America

Oscar Schofield: Carbon dynamics on the Mid-Atlantic Bight: Sources and sinks seen from space

Coastal process modeling
moderated by Niki Gruber, Galen McKinley

Brief science presentations:

Amala Mahadevan: Wind-driven exchange of carbon at the continental shelfbreak.

Sybil Seitzinger: Modeling DOM inputs to coastal systems from watersheds (watershed, regional and global perspectives).

Debby Ianson: Carbon fluxes in coastal upwelling and relaxation regions: Sources or sinks?

Niki Gruber: Influence of mesoscale processes on coastal productivity and air-sea CO_2 fluxes.

End Day II @ 18:00

Day III, 23 September:

Session IIIA: Plenary: Breakout group summary, beginning at 0800.

08:00-09:00   Plenary: Brief summary presentations from Breakout Groups 3

Session IIIB, 09:00-12:30: Breakout Groups 4: Crosscutting themes and synthesis

09:10-12:30   Breakout groups 4:
Cross-cutting/synthesis group A: Boundaries
Cross-cutting/synthesis group B: Pre-Synthesis
Cross-cutting/synthesis group C: Process Study

Session IIIC. 13:30-14:30: Plenary; summary reports from Breakout Groups 4, beginning at 13:30

13:30-14:30   Plenary: Brief summary presentations from Breakout Groups 4

Session IIID: Plenary; Summary presentation: Inter-agency support (by PMs)

Workshop end

References

Denning, A.S., et al. 2005. *Science Implementation Strategy for the North American Carbon Program.* Report of the NACP Implementation Strategy Group of the US Carbon Cycle Interagency Working Group. US Carbon Cycle Science Program, Washington, DC, 68 pp.

Doney, S.C., R. Anderson, J. Bishop, K. Caldeira, C. Carlson, M.-E. Carr, R. Feely, M. Hood, C. Hopkinson, R. Jahnke, D. Karl, J. Kleypas, C. Lee, R. Letelier, C. McClain, C. Sabine, J. Sarmiento, B. Stephens, and R. Weller, 2004. *Ocean Carbon and Climate Change (OCCC): An Implementation Strategy for U. S. Ocean Carbon Cycle Science.* UCAR, Boulder, CO, 108 pp. Available at http://www.carboncyclescience.gov.

Wofsy, S.C., and R.C. Harriss, 2002. *The North American Carbon Program (NACP).* Report of the NACP Committee of the U.S. Interagency Carbon Cycle Science Program. US Global Change Research Program, Washington, DC, 75 pp.

North America's Atlantic Coast

Rick Jahnke
Skidaway Institution of Oceanography

Oscar Schofield
Institute of Marine and Coastal Sciences
Rutgers University

Wei-Jun Cai
University of Georgia

Introduction

This passive margin stretching from Florida to Newfoundland consists of broad, shallow, sandy shelf areas inshore of the Gulf Stream (Figure 3.1). Significant wind-driven upwelling does not occur here and water exchange with the open ocean takes place through mesoscale eddies and meandering filaments. North Atlantic open-ocean nutrient levels are relatively low. Seasonal thermal forcing is intense. Extreme storm events are significant factors affecting stratification, nutrient distributions, productivity, and sediment transport (Figures 3.2-3.4). Terrestrial and riverine inputs are important, and their character and impact varies latitudinally. To the north, rivers entering the ocean via large estuaries/embayments dominate; to the south, salt marshes are the primary site of land-ocean exchange (Figure 3.5). Anthropogenic activities on this coastline are intense, with close to 30% of the US population living within 150 km of the coast.. This coast's impact on airmasses is seasonally variable. In winter, there is mean offshore atmospheric transport across most of this coastline. In summer, offshore transport is largely relegated to the Labrador coast, while there is mean onshore transport in the South Atlantic and Mid Atlantic Bights.

Subregions of the Atlantic Coast

The East Coast appears to consist of at least three distinct subregions, the South- and Mid-Atlantic Bights (SAB and MAB, respectively), and the Scotian shelf/ Gulf of Maine system, which are all distinct with regard to their circulation patterns, atmospheric forcing, productivity, anthropogenic impacts, and interactions with adjacent terrestrial systems.

The South Atlantic Bight

The coast of the SAB, which extends from the northern Florida Straits to Cape Hatteras, is characterized by extensive salt marshes, which account for 80% of all intertidal marshes along the north Atlantic coast (Figure 3.5). There is little riverine freshwater input to the coastal ocean through these marshes. At the seaward boundary of the SAB, the Gulf Stream lies right on the shelf break, separating the SAB from the open North Atlantic. In the isolated waters between shore and shelf break, waters experience intense heating, but little seasonality in insolation

Figure 3.1. Map of the US Atlantic coast, showing the predominance of sandy sediments.

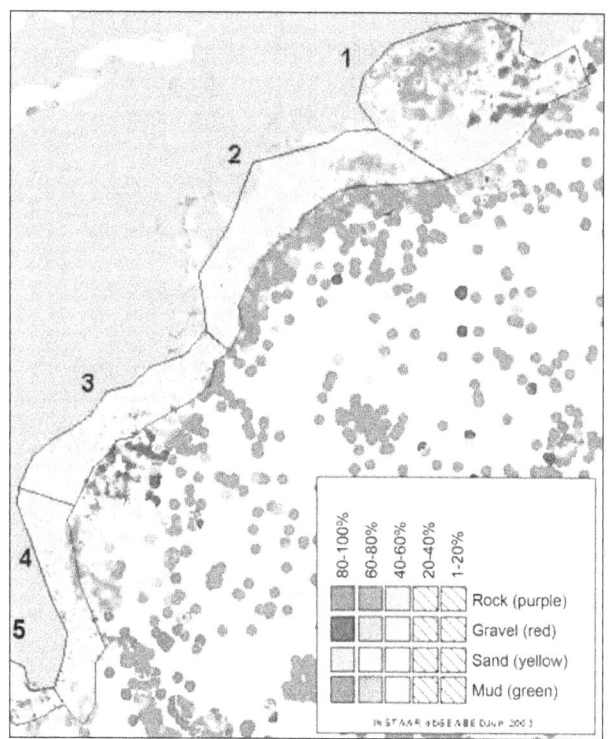

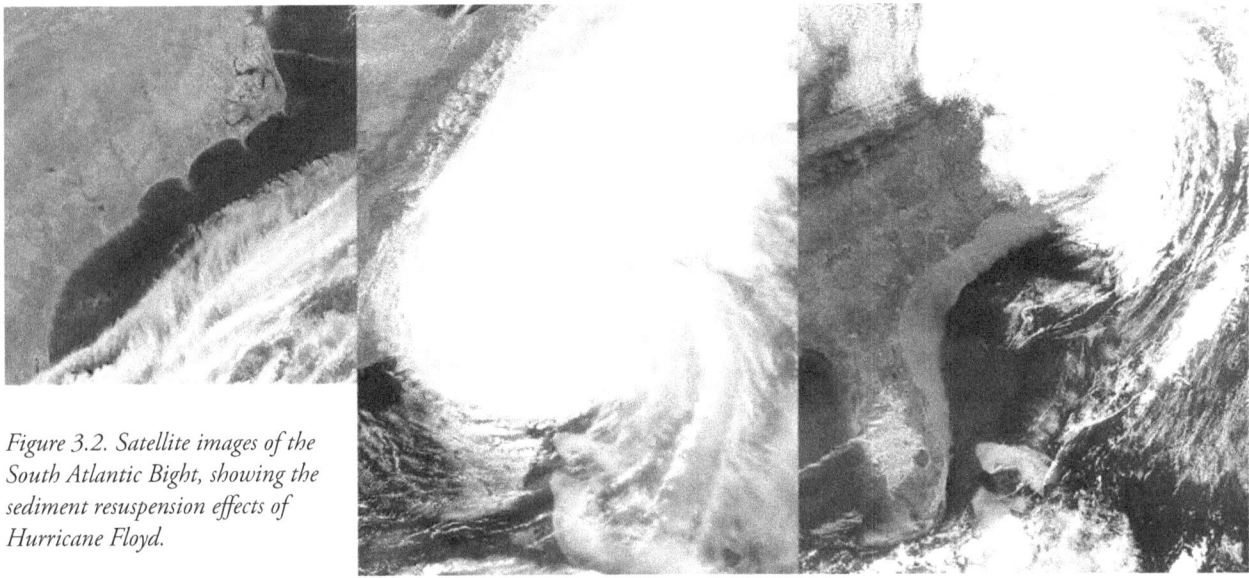

Figure 3.2. Satellite images of the South Atlantic Bight, showing the sediment resuspension effects of Hurricane Floyd.

forcing (Yoder et al. 1981). Despite the strong surface heating, water-column stratification in the SAB is weak throughout the year. Open-ocean nutrients are low, and anthropogenic impacts are comparatively low on the SAB. The SAB shelf environment is also shaped in large part by episodic atmospheric processes (storms) from the convective scale to continental scale, impacting buoyancy and momentum exchanges, inputs of nutrients and trace metals through both wet and dry deposition, as well as water vapor and gas fluxes. This ensemble of processes extends through a great range of temporal scales and spatial patterns, from seconds and centimeters (e.g., turbulent fluctuations and transports) to many years and hundreds to thousands of kilometers (e.g., ocean regime shifts and climate change).

The SAB is the longest stretch of our national coastal ocean that is affected persistently by a western boundary current. The Gulf Stream has first-order effects on the circulation and hydrography of much of the SAB continental shelf and slope. It plays an important, sometimes dominant role in determining the exchanges of heat, momentum, nutrients, inorganic compounds, and biological constituents across the continental shelf-deep ocean interface. Intrusions driven by Gulf Stream meanders are the main source of nutrients to the SAB shelf (Jahnke and Blanton, in press), estimated to be greater than 9×10^{10} mol N yr^{-1} (Lee et al. 1991). Gulf Stream meanders and their attendant cyclonic,

Figure 3.3. High-resolution cross-shelf sections across the shelf at New Jersey, using a Webb Glider. Note the strong resuspension signature in the mid-shelf as a result of Hurricane Ivan.

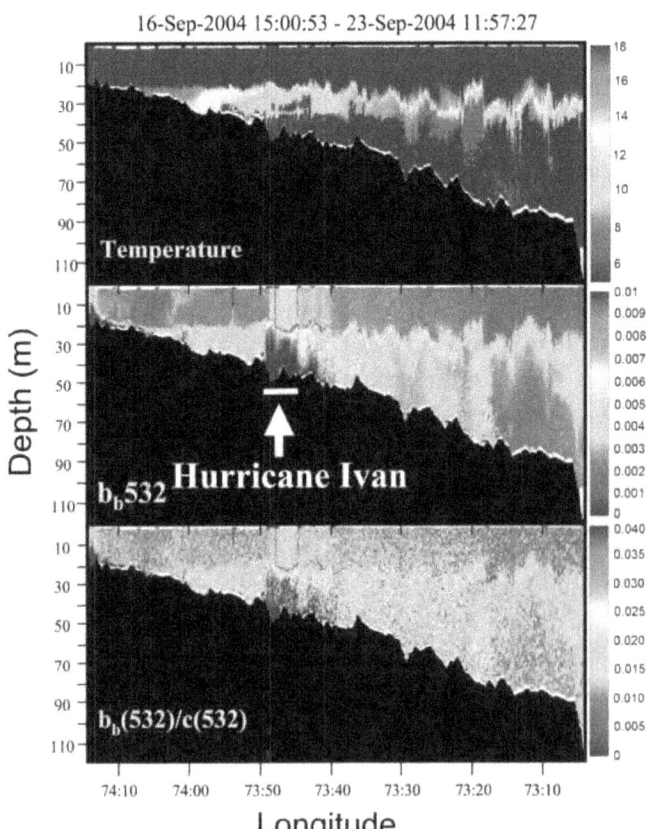

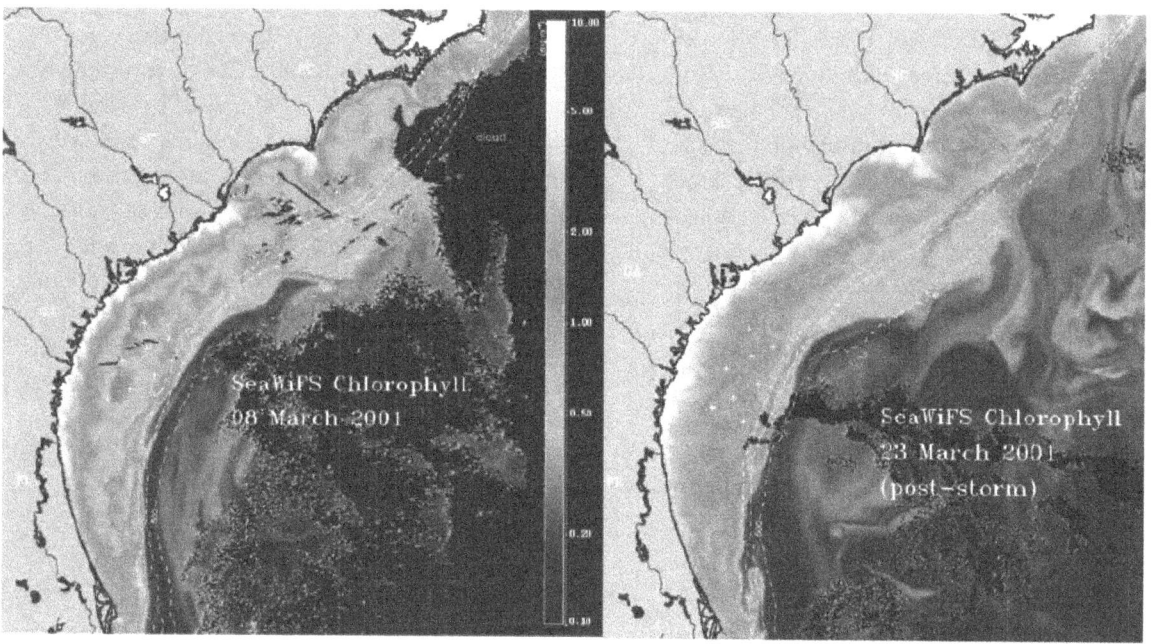

Figure 3.4. SeaWIFS image of the South Atlantic Bight, showing the productivity response to a winter storm.

cold-core frontal eddies form at 2- to 14-day intervals within and north of the Florida Straits, propagate northward along the shoreward edge of the Gulf Stream, encountering regions of growth and decay from the Straits until the Gulf Stream separates at Cape Hatteras (Bane et al. 1981; Lee et al. 1991). The extent to which meander-driven intrusions penetrate onto the shelf depends directly on shelf hydrography, which is largely determined by seasonal atmospheric conditions (Atkinson et al. 1983). Under northward wind stress and summer stratification, subsurface intrusions can occupy almost the entire width of the shelf (e.g., Lee and Pietrafesa, 1987; Aretxabaleta et al. 2005). From late fall to early spring, intrusions are less likely to penetrate to the mid-shelf and benthic processes may become more important. The degree of penetration and fate of meanders is not well known beyond this coarse seasonal breakdown. It is critical to understand the temporal and spatial variability of this process in the SAB, and since nutrient input/export, new production, and carbon export from the shelf are strongly determined by the meander decay and growth regions, these link directly to the Gulf Stream.

The Mid Atlantic Bight

The coast of the MAB is dominated by large estuaries or embayments, such as Chesapeake Bay, Delaware Bay, the New York/New Jersey Apex (Hudson River estuary), and Long Island Sound. The Gulf Stream begins to separate from the coastline as the shelf broadens at about Cape Hatteras, and the shelf of the MAB is overall broader than that of the SAB. Unlike the SAB, the seaward boundary of this region is not the Gulf Stream, but a front that persists at

Figure 3.5. Geographic distribution of salt marshes on the US East Coast, is based on early data. Marsh areas in the northeast states have declined significantly over last few decades.

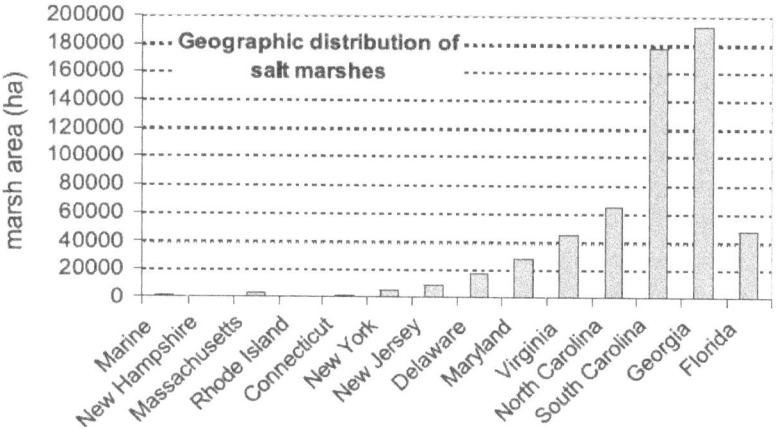

the shelf break. The MAB shows typical mid-latitude seasonality in insolation forcing, with dramatic warming of summertime coastal waters relative to winter conditions. Along with this insolation forcing comes a seasonal change in stratification: MAB waters move from essentially unstratified in winter to strongly stratified by late summer.

The large-scale circulation in the MAB is characterized as an alongshore flow toward the south at about 5 cm s^{-1} (Beardsley and Boicourt, 1981). This current transports fresher water along the shelf from the Labrador Sea to Cape Hatteras (Chapman and Beardsley, 1989); on any given day, however, the mean shelf flow is rarely observed. Short-term wind forcing with typical time scales of one to three days often results in much stronger currents (typically 20-40 cm s^{-1}) that flow with the wind in either alongshelf direction (Beardsley and Boicourt, 1981) and are clearly steered by nearshore topographic features (Kohut et al. 2004). As the large-scale shelf current flows southward, it is subject to the cascading impacts of freshwater inputs from the numerous urbanized rivers and bays on the inshore side. This integrated plume responds to strong downwelling winds by flowing alongshelf to the south along New Jersey, and similarly responds to strong upwelling winds by flowing alongshelf to the east along Long Island. During weak winds, much of the Hudson plume is observed flowing along an alternate cross-shelf pathway following the southern flank of the Hudson shelf valley and extending to the mid- or outer shelf (Glenn et al. 2004) and accounts for up to 70% of the low-salinity water found on the outer edge of the shelf. In late spring and early summer, a strong thermocline develops at about 20 m depth across the entire shelf, isolating a continuous mid-shelf cold pool of water that extends from Nantucket to Cape Hatteras (Houghton et al. 1982).

On the offshore side of the shelf, the southward flowing shelf current has been likened to a leaky hose that continuously loses freshwater across the shelf break to the continental slope (Loder at al. 1998). The exchange processes, however, are complicated by the shelf-break front that separates the relatively cool, fresh shelf waters from warmer, saltier slope water (Iselin, 1936; Houghton et al. 1988). The front is a highly variable narrow jet with time scales of the order of a day and length scales of the order of 10 km

(Linder and Gawarkiewicz, 1998; Fratantoni et al. 2001; Garwarkiewicz et al. 2001, 2004). It is thought that cross-frontal exchange is dominated by wind forcing during the winter (Lozier and Garwarkiewicz, 2001) and by frontal instabilities during the summer (Houghton et al. 1988), although warm core Gulf Stream rings that propagate along the shelf break contribute an unknown but likely significant amount to transports across the shelf break (see Figure 3.6). Climatological temperature and salinity observations (Linder et al. 2004) indicate the front is typically about 20 km wide, and extends from surface to bottom with the foot of the front along the 80 m isobath. The foot of the front is often noted by a turbidity maximum (Barth et al. 1998) where the boundary layer detaches, flowing along isopycnals offshore and toward the surface at speeds of about 10 m d^{-1} (Houghton, 1997; Houghton and Visbeck, 1998). It is thought that this secondary circulation may carry nutrients into the euphotic zone resulting in enhanced phytoplankton and a second strong turbid alongshore front at the shelf break (Ullman and Codiga, 2004).

The Scotian Shelf and Gulf of Maine

The Gulf of Maine (GoM), located on the North American continental shelf between Cape Cod and Nova Scotia, is a semi-enclosed basin opening to the North Atlantic Ocean. The geometry of the GoM is dominated by several deep basins and shallow submarine banks. On the seaward flank of the GoM is Georges Bank (GB), which is separated from the Nantucket Shoals to the west by the Great South Channel (GSC) and from the Scotian Shelf to the east by the Northeast Channel. The first systematic study of the general circulation in the GoM can be traced back to Bigelow (1927) who, using surface drifters and hydrographic surveys, suggested that the summertime (stratified season) surface circulation in the GoM is dominated by two relatively large-scale gyres: a cyclonic circulation around Jordan Basin and an anticyclonic circulation around Georges Bank. This was confirmed by long-term direct Eulerian and Lagrangian current measurements. Flushing of the GoM occurs through well-defined channels and there is a persistent buoyancy-driven equatorward coastal current that is driven by a distributed river system.

Figure 3.6. Schematic depiction of potential cross-shelf transport pathways in the MAB.

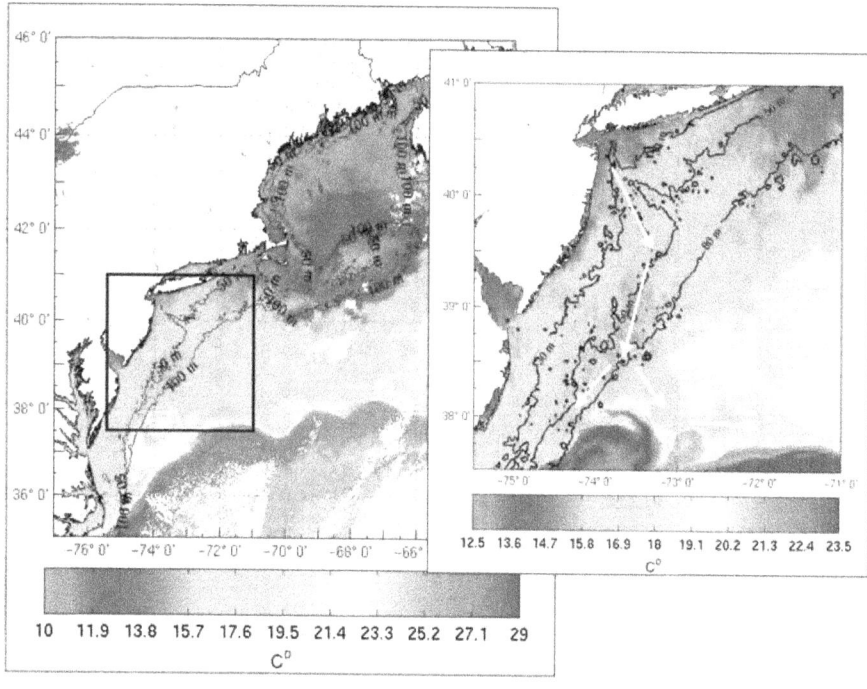

The Atlantic Coast's Carbon Cycle

In the SAB, low supply rates of terrestrial and open-ocean nutrients lead to low overall water column productivity. Large storms may be significant factors supplying nutrients to the coastal ocean, as evidenced by the observations of phytoplankton blooms following hurricanes (Figure 3.4). Low water column productivity and shallower shelf depths allow light penetration to the seafloor on much of the East Coast, suggesting that benthic photosynthesis may be important (Figure 3.7). Indeed, preliminary calculations suggest that benthic photosynthesis contributes up to one-third of the total new productivity on the mid-shelf. Benthic burial and respiration can only account for about half of water column and benthic productivity; it is uncertain as to whether the difference is accounted for by water-column respiration or off-shelf transport of C. A preliminary carbon budget exercise suggests that the shelf-wide respiration rate may be slightly higher than the primary production rate. This net heterotrophy may be subsidized, or even offset, by the respiration of marsh- and river-exported organic carbon (Cai et al. 2003; Figure 4.8). Large storm events may be significant transporters of sedimentary C off the shelf to the adjacent deep ocean (Figure 3.2). In the SAB, the inner and outer shelves are more productive than

the mid-shelf. The annual average was estimated as 386 $gC\ m^{-2}\ yr^{-1}$ (or 32 mol C $m^{-2}\ yr^{-1}$ or 1.1 $gC\ m^{-2}\ day^{-1}$) from available nutrient flux from the subsurface Gulf Stream water (Yoder et al. 1985; Li et al. 1991; Menzel et al. 1993). Dark incubation data suggest that SAB has a higher respiration rate than the primary productivity (PP, Menzel et al. 1993).

Net productivity in the MAB is elevated compared to the SAB, probably due in part to larger supplies of terrestrially derived nutrients (both natural and anthropogenic), the higher nutrient content of the open Atlantic at these higher latitudes, and the stratification and secondary circulation that result from the terrestrial fresh water inputs. In addition, although direct measurement of respiration rate is not available for the MAB, it is most likely less than the PP (Kemp, 1995). The MAB experiences a spring phytoplankton bloom, that begins in the upper water column. As surface nutrients are depleted, the phytoplankton form subsurface particle maxima near the pycnocline. The nearshore side of the subsurface cold pool is forced by wind-driven Ekman transport in the surface layer, causing cycles of coastal upwelling and downwelling that fuel 25% of the shelf phytoplankton productivity (Glenn et al. 2004). The nearshore blooms are advected offshore in squirts to mid-shelf (Bosch et al. 2004), stalling at the midshelf front located along the 60 m

Estimated PAR Reaching the SAB Shelf Sea Floor (PCT Surface)

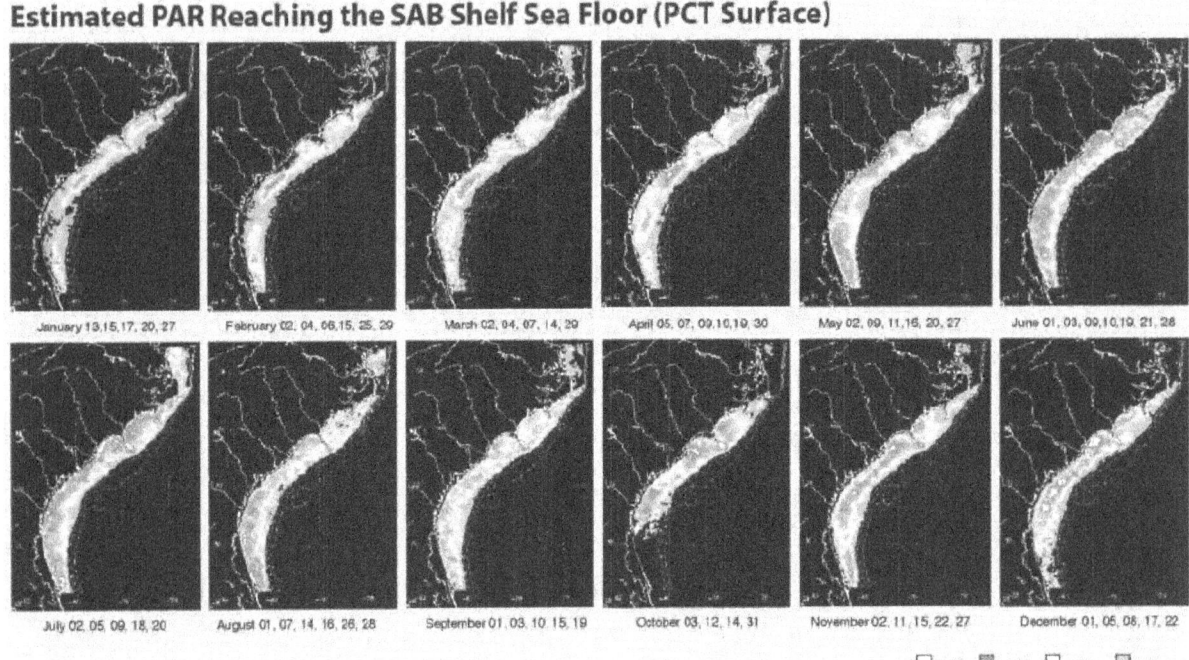

January 13,15,17, 20, 27   February 02, 04, 06,15, 25, 29   March 02, 04, 07, 14, 29   April 05, 07, 09,10,19, 30   May 02, 09, 11,16, 20, 27   June 01, 03, 09,10,19, 21, 28

July 02, 05, 09, 18, 20   August 01, 07, 14, 16, 26, 28   September 01, 03, 10, 15, 19   October 03, 12, 14, 31   November 02, 11, 15, 22, 27   December 01, 05, 08, 17, 22

≥15% ≥10% ≥5% ≥1%

Figure 3.7. The impact of benthic photosynthesis on the carbon cycle in the SAB.

Summary of Primary Production and Benthic Respiration on the SAB Shelf (gC m⁻² yr⁻¹)

Primary Production

Water Column:	248
Benthic:	91 - 128
Total Production:	339 - 376

Benthic Respiration

Interface:	91 - 128
Sedimentary:	50
Total Benthic Respiration:	141 - 178

Sediment Respiration: Total Production   0.38 - 0.53

isobath (Ullman and Codiga, 2004). Following the spring bloom, MAB primary productivity decreases with a summer bloom of moderate rates, and relatively low rates during winter. The southern part of the MAB also has a higher PP than the northern part (Lohrenz et al. 2002; Verity et al. 2002; O'Reilly and Busch, 1984; Falkowski et al. 1983; Malone et al. 1983). An annual average PP for the MAB is around 290 to 365 gC m⁻² yr⁻¹ (0.8-1.0 gCm⁻² day⁻¹).

The productivity represents the contributions of rivers, coastal upwelling, and shelf productivity. The chlorophyll associated with rivers and coastal upwelling each represent 25% of the annual MAB phytoplankton biomass (Figure 3.9). As elsewhere along this coast, benthic diagenesis and sediment burial of carbon

cannot account for a large fraction of the organic carbon delivered from rivers or produced in the water column, and it is uncertain if the excess is accounted for by water-column respiration or export to the ocean interior. Many of the unresolved budgets reflect complex circulation pathways on the MAB driven by local wind forcing, the Hudson canyon, which appears to play an important role in transporting nearshore carbon offshore, and offshore warm core rings (Figure 3.9). Although hurricanes do affect this portion of the coast, they are not as common as they are in the SAB; in contrast, winter storms are more common here than in the SAB.

The carbon cycle in the GoM/Scotian Shelf is dominated by spring and summer blooms fueled by nutrients derived from both terrestrial and oceanic sources that are photosynthetically consumed when increases in insolation and decreases in winter storm conditions allow the water column to stratify sufficiently to alleviate light limitation. The productivity and phytoplankton community composition is impacted by climate variability, which can change nutrient availability by altering waters entering the GoM.

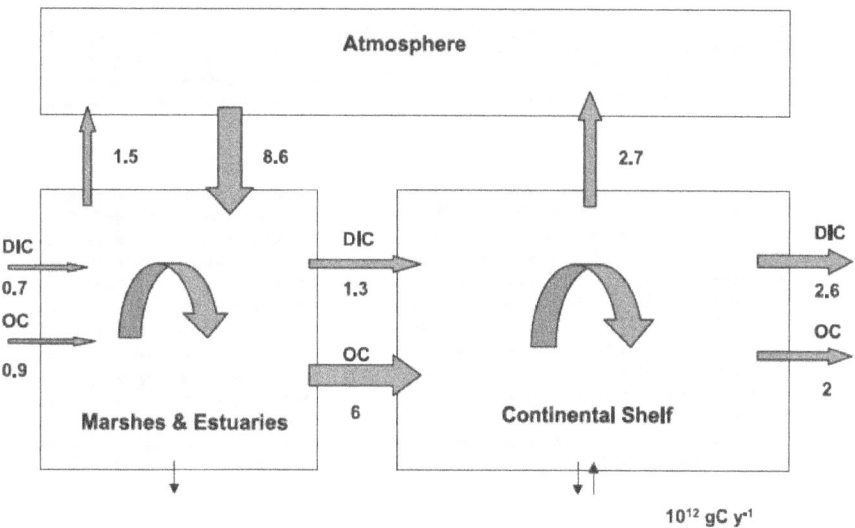

Figure 3.8. A schematic carbon budget for the SAB, including net primary productivity in the salt marshes.

Figure 3.9. The contribution of riverine, shelf-water, and upwelling-derived nutrients to primary productivity in the MAB. Water mass classification algorithms of SeaWIFS ocean color imagery allow distinction of the chlorophyll associated with rivers and coastal upwelling. The remaining chlorophyll represents general shelf productivity. Rivers and coastal upwelling each account for ~25% of the total MAB chlorophyll a.

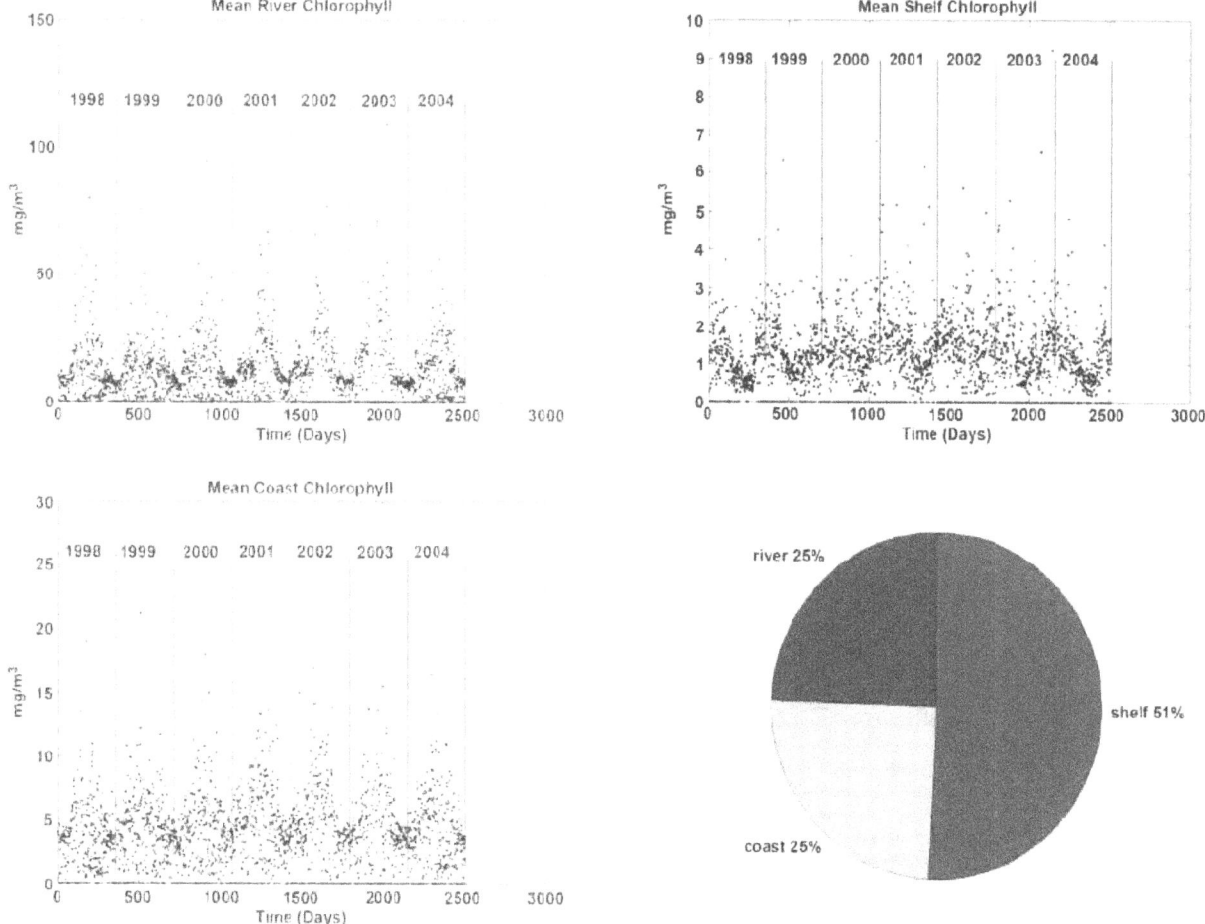

Atlantic Coast Air-Sea CO₂ Exchange

Although there have been only a few studies dedicated to air-sea exchange of CO_2 on the East Coast, net transfer of CO_2 seems to be highly latitudinally dependent. Waters of the central SAB appear to be an annual source of CO_2 to the atmosphere (Figure 3.10; Cai et al. 2003), the result of a large export of inorganic and organic carbon from intertidal marshes, low overall water-column net productivity, and strong heating. Recent work shows that while surface water pCO_2 in the SAB shows great variability both spatially and temporally, the annual flux in the whole SAB shelf area is nearly zero (Jiang et al. submitted). Recognizing the significant C export from the adjacent salt marshes, these observations imply a significant net removal of CO_2 from the coastal atmosphere. Waters of the MAB seem to be a large annual sink of atmospheric CO_2 (Figure 3.11; Degrandpre

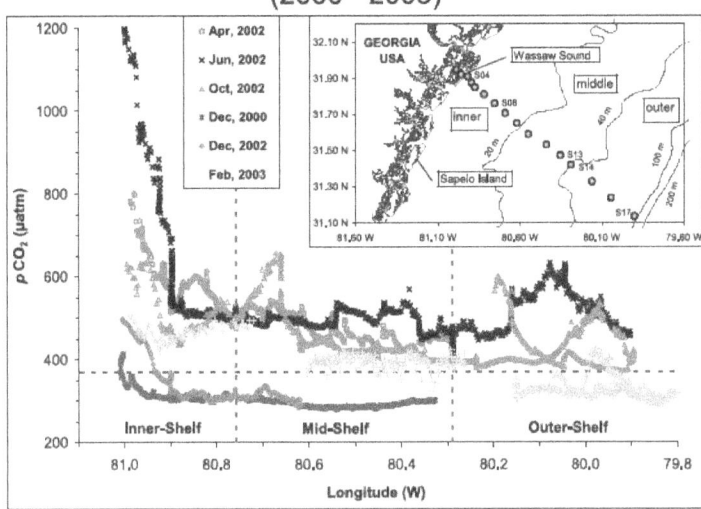

Surface pCO_2 in the South Atlantic Bight (2000 - 2003)

Figure 3.10. Surface pCO₂ distributions in the SAB. From Cai et al. (2003).

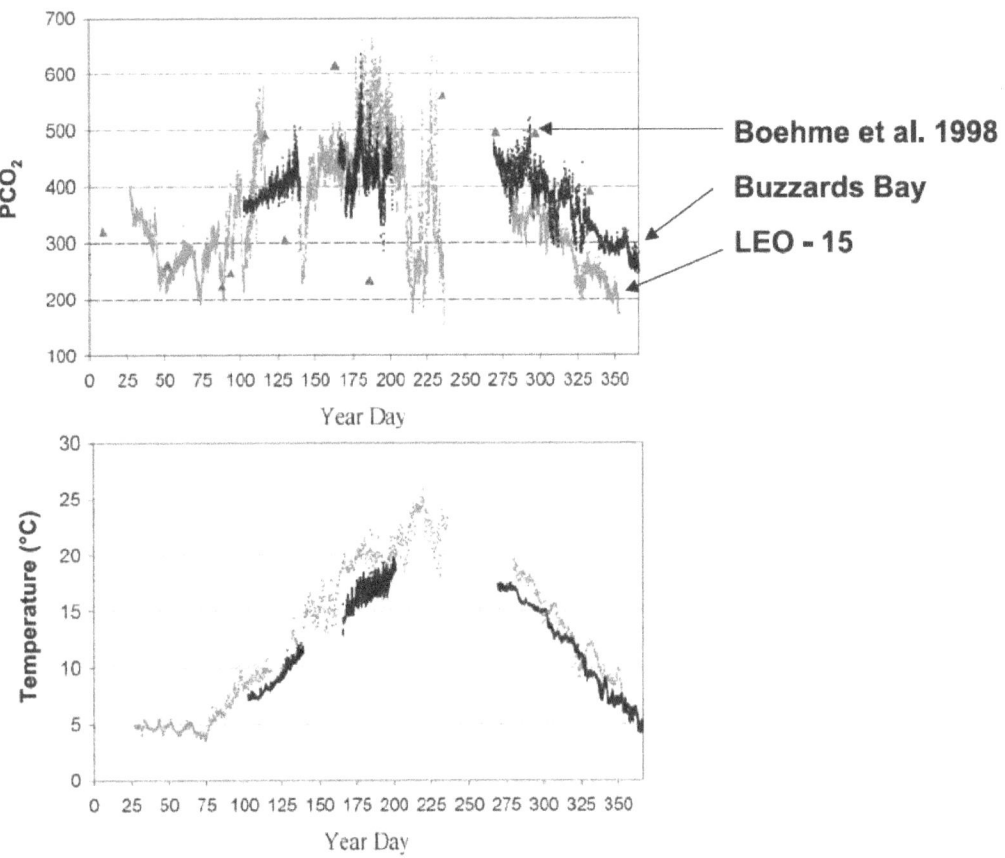

Figure 3.11. Surface pCO₂ distributions in the MAB. From DeGrandpre et al. (2002).

SABSOON
offshore Navy towers

ATMOSPHERE MEASUREMENTS:
Wind speed & direction
Air temperature, humidity, and barometric pressure
Short & Long wave and photosynthetically available radiation
Precipitation

OCEAN MEASUREMENTS:
Sea-surface temperature
Waves (non-directional), water level
Temperature and Conductivity (salinity)
Chlorophyll fluorescence
ADCP Current Profiles

DATA TELEMETRY:
2-way real-time microwave link to shore
T1 landline to SkIO

Locations of TACTS towers and sensor configurations.

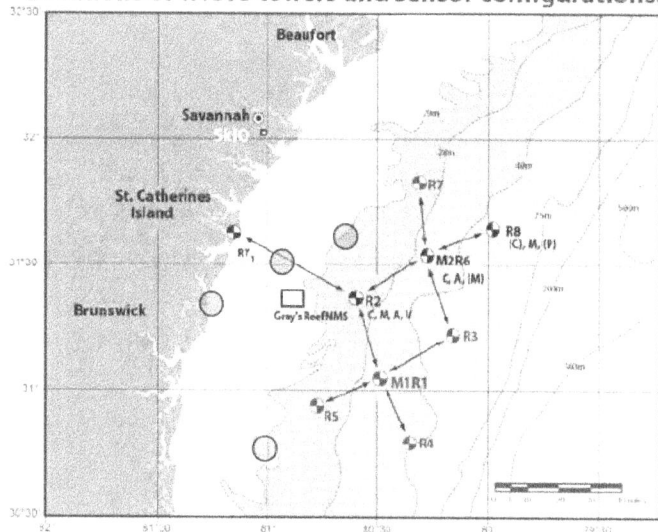

Figure 3.12. Observatory systems in the SAB.

et al. 2002), probably due to relatively greater inputs of nutrients that fuel summer productivity capable of negating the effects of annual heating. The Go appears to be a sink of atmospheric CO_2 (Vandemark et al. 2006).

Research Programs in Atlantic Coastal Waters

There have been several large historical measurement programs relevant to the study of carbon cycling in the waters of the East Coast. These include the Shelf-Edge Exchange Program (SEEP), the Department of Energy's Ocean Margins Program (DOE-OMP), and the NSF Langragian Transport and Transformation Experiment (LATTE). Future programs will be able find significant leverage as the East Coast has one of the most extensive arrays of ocean observing networks in the world, which offers the potential infrastructure to support extensive carbon cycling programs. Of particular note are the electro-optical cables that provide the high power capabilities

to support sensors central to carbon biogeochemistry. Beginning in the SAB, the SABSOON array on the inshore side of the Gulf Stream (Figure 3.12) is an observatory consisting of a series of towers that provide an excellent platforms for both ocean and atmospheric measurements within the footprint of shore-based HF radar current and wave measurements. In the MAB, the Long term Ecosystem Observatory (LEO) was deployed in 1996. It consists of electro-optical cable 5 km offshore in 15 m of water outfitted with a vertical profiler. This site is embedded within the larger New Jersey Shelf Observing System consisting of nested HF radar networks, and sustained subsurface Gliders. Finally, on the northern edge of the MAB is the Martha's Vineyard cabled observatory, the world's most advanced cable system. It ends in a tower system allowing researchers to make both in-water and atmospheric measurements (Figure 3.13).

Anthropogenic Impacts on the Atlantic Coast

The Northeast has experienced strong anthropogenic forcing given the heavy urbanization along the coasts. For example, the Hudson buoyant plume has maximum discharge rates on the order of 1000 m^3 s^{-1}, which is not large relative to the Mississippi or Columbia Rivers. Despite the modest flow rates, the Hudson dominates the transport of nutrients and chemical contaminants to the coastal

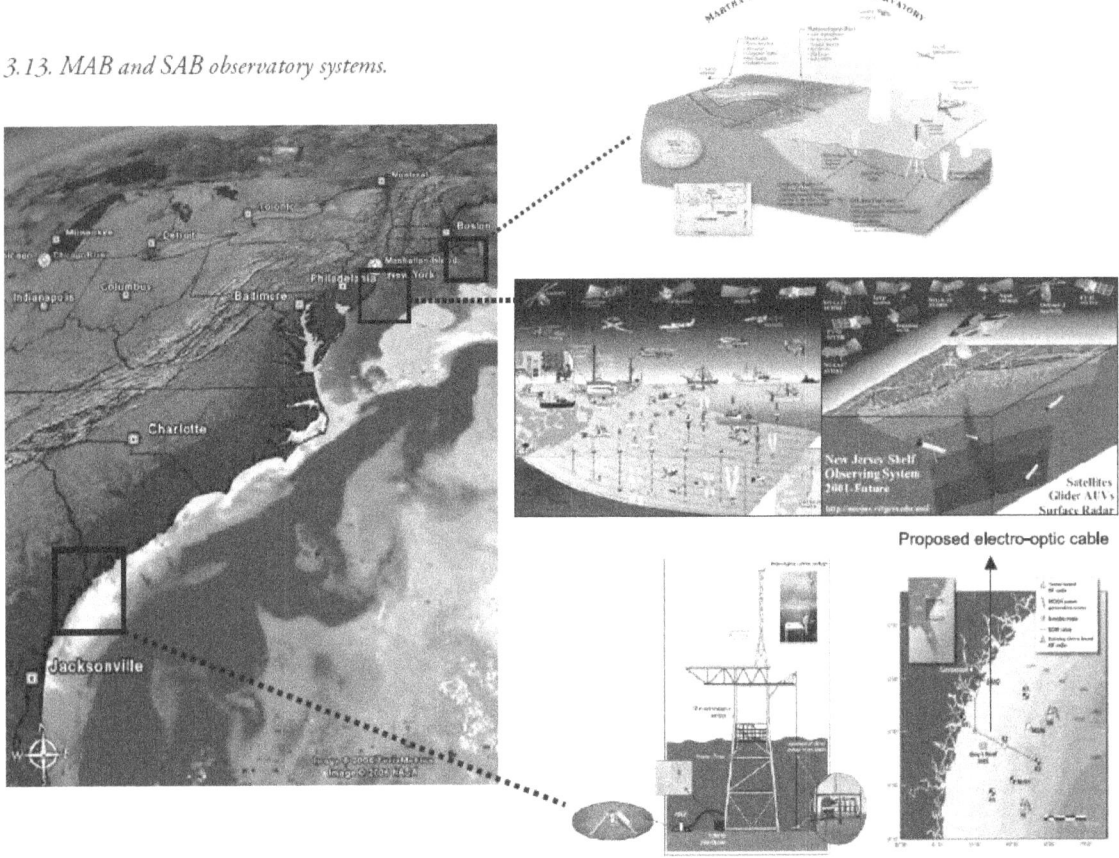

Figure 3.13. MAB and SAB observatory systems.

Proposed electro-optic cable

waters of the MAB, as for over 100 years it has been the most urbanized estuary in the United States. For example, only recently has Los Angeles' population exceeded what New York's was in 1900, and today over 20 million people live in the Hudson River watershed (Gibson, 1998). Currently 100 m³ s⁻¹ of treated sewage flow into the lower estuary (ISC, 1997) and 90% of the associated inorganic nitrogen is exported unassimilated to the coastal ocean (Garfield, 1976). This is in contrast to the Chesapeake where most nitrogen discharged into the estuary is assimilated by phytoplankton within the estuary (Malone et al. 1996) and exported to the coastal ocean as organic nitrogen. This provides significant nitrogen to coastal waters and contributes to the majority of 25% of chlorophyll associated with rivers in the MAB (Figure 3.9).

In addition to these direct impacts of human activities, the ecosystems in the Northwest Atlantic are particularly sensitive to climate change (Frank et al. 1990). The continental shelf in this region is located in a faunal transition zone, with Arctic-boreal, temperate, and subtropical/tropical species boundaries in close proximity (Herman, 1979). With its large annual cycle of water temperature, the Northwest Atlantic continental shelf is both the northern limit for many warm water organisms and the southern limit for many cold water species. The trend toward warmer MAB shelf waters could cause a significant latitudinal shift in the faunal transition zone and thus in the seasonal occurrence of different organisms in the study area (Greve et al. 2001). Durbin and Durbin (1996) have predicted that changes in water column temperature due to climate warming will have maximum impact on the population dynamics of planktonic species that predominate in the wintertime. Such changes are already occurring on the continental shelf areas of the eastern Atlantic (Reid et al. 2001; Beaugrand et al. 2002).

References

Aretxabaleta, A., J. Manning, F.E. Werner, K. Smith, B.O. Blanton, and D.R. Lynch, 2005. Data assimilative hindcast on the southern flank of Georges Bank during May 1999: Frontal circulation and implications. *Continental Shelf Research*, 25:849-874.

Atkinson, L.P., T.N. Lee, J.O. Blanton, and W.S. Chandler, 1983. Climatology of the southeastern United States continental shelf waters. *Journal of Geophysical Research*, 88:4705-4747.

Bane, J.M., D.A. Brooks, and K.R. Lorenson, 1981. Synoptic observations of the three-dimensional structure and propagation of Gulf Stream meanders along the Carolina continental margin. *Journal of Geophysical Research*, 86:6411-6425.

Barth, J.A., D. Bogucki, S. Pierce, and P. Kosro, 1998. Secondary circulation associated with a shelfbreak front. *Geophysical Research Letters*, 25:2761-2764.

Beardsley, R.C., and W.C. Boicourt, 1981. On estuarine and continental shelf circulation in the Middle Atlantic Bight. In: *Evolution of Physical Oceanography* [B.A. Warren and C. Wunsch (eds.)]. MIT Press, Cambridge, MA.

Beaugrand, G., P.C. Reid, F. Ibañez, J.A. Lindley, and M. Edwards, 2002. Reorganization of North Atlantic marine copepod biodiversity and climate. *Science*, 296:1692-1694.

Bigelow, H.B., 1927. Physical oceanography of the Gulf of Maine. *Bulletin of the Bureau of Fisheries*, 40:511-1027.

Bosch, J., O. Schofield, S. Glenn, and J. Kohut, 2004. East coast plumes and blooms: Building a chlorophyll budget for the Mid Atlantic Bight. *Ocean Optics XVII*, 25-29 Oct 2004, Fremantle, Australia.

Cai, W.J., Z.A. Wang, and Y. Wang, 2003. The role of marsh-dominated heterotrophic continental margins in transport of CO_2 between the atmosphere, the land-sea interface and the ocean. *Geophysical Research Letters*, 30:1849, doi:10.1029/2003GL017633.

Chapman, D., and R. Beardsley, 1989. On the origin of shelf water in the Middle Atlantic Bight. *Journal of Physical Oceanography*, 19:384-391.

DeGrandpre, M.D., T.R. Hammar, G.J. Olbu, and C.M. Beatty, 2002. Air-sea CO_2 fluxes on the US Middle Atlantic Bight. *Deep-Sea Research II*, 49:4355-4367.

Durbin, E.G., and A.G. Durbin, 1996. Zooplankton dynamics in the northeast shelf ecosystem. In: *The Northeast Shelf Ecosystem: Assessment, Sustainability, and Management*. Blackwell Science, Inc., p. 564.

Falkowski, P.G., J. Vidal, T.S. Hopkins, G.T. Rowe, T.E. Whitledge, and W.G. Harrison, 1983. Summer nutrient dynamics in the Middle Atlantic Bight: Primary productivity and the utilization of phytoplankton carbon. *Journal of Plankton Research*, 5:515-537.

Frank, K., R.I. Perry, and K.F. Drinkwater, 1990. Predicted response of northwest Atlantic invertebrate and fish stocks to CO_2 induced climate change. *Transactions of the American Fisheries Society*, 119:353-365.

Fratantoni, P., R. Pickart, D.J. Torres, and A. Scotti, 2001. Mean structure and dynamics of the shelfbreak jet in the Middle Atlantic Bight during fall and winter. *Journal of Physical Oceanography*, 31:2135-2156.

Garfield, E., 1976. Current comments: ISI's chemical information system goes marching on. *Current Comments*, 1 (reprinted in *Essays of an Information Scientist*, 2:402-403).

Garwarkiewicz, G.F., F. Bahr, R.C. Beardsley, and K.H. Brink, 2001. Interaction of a slope eddy with the shelfbreak front in the Middle Atlantic Bight. *Journal of Physical Oceanography*, 31:2783-2796.

Garwarkiewicz, G., K.H. Brink, F. Bahr, R.C. Beardsley, M. Caruso, J.F. Lynch, and C.-S. Chiu, 2004. A large-amplitude meander of the shelfbreak front in the Middle Atlantic Bight: Observations from the Shelfbreak PRIMER experiment. *Journal of Geophysical Research*, 109:C03006, doi:10.1029/2002JC001468.

Gibson, C., 1998. *Population of the 100 Largest Cities and Other Urban Places in the United States: 1790 to 1990*. Campbell Gibson Population Division, U.S. Bureau of the Census, Washington, DC.

Glenn, S., R.A. Arnone, et al. 2004. The biogeochemical impact of summertime coastal upwelling on the New Jersey Shelf. *Journal of Geophysical Research*, 109:C12S02, doi:10.1029/2003JC002265.

Greve, W., U. Lange, et al. (2001). Predicting the seasonality of the North Sea zooplankton. *Senckenbergiana maritima*, 31:263-268.

Herman, Y., 1979. Plankton distribution in the past. In: *Zoogeography and Diversity of Plankton* [S. Van Der Spoel and A.C. Pierrot-Bults (eds.)]. John Wiley & Sons, New York, pp. 29-49.

Houghton, R.W., 1997. Lagrangian flow at the foot of a shelfbreak front using a dye tracer injected into the bottom boundary layer. *Geophysical Research Letters*, 24:2035-2038.

Houghton, R.W., and M. Visbeck, 1998. Upwelling and convergence in the Middle Atlantic Bight shelfbreak front. *Geophysical Research Letters*, 25:2765-2768.

Houghton, R.W., R. Schlitz, R.C. Beardsley, B. Butman, and J.L. Chamberlin, 1982. The Middle Atlantic Bight Cold Pool: Evolution of the temperature structure during summer 1979. *Journal of Physical Oceanography*, 12:1019-1029.

Houghton, R.W., F. Aikman III, and H.W. Ou, 1988. Shelf-slope frontal structure and cross-shelf exchange at the New England shelf-break. *Continental Shelf Research*, 8:687-710.

ISC (Interstate Sanitation Commission), 1997. *Annual Report of the Interstate Sanitation Commission on the Water Pollution Control Activities and the Interstate Air Pollution Program.* New York State Document SAN 750-178-50638.

Iselin, C.O., 1936. A study of the circulation of the western North Atlantic. *Papers in Physical Oceanography and Meteorology*, 4:1-101.

Jahnke, R.A. and J.O. Blanton, in press. Western Boundary Currents - The Gulf Stream. In: *Carbon and Nutrient Fluxes in Continental Margins: A Global Synthesis* [K.K. Liu, L. Atkinson, R. Quinones, and L. Talaue-McManus (eds.)] Global Change: The IGBP Series, Springer-Verlag.

Jiang, L.-Q., W.-J. Cai, Y. Wang, R. Wanninkhof, and H. Lüger. Air-sea CO_2 fluxes on the US South Atlantic Bight. Submitted to *Journal of Geophysical Research-Oceans*.

Kemp, P., 1995. Can we estimate bacterial growth rates from ribosomal RNA content? *Molecular Ecology of Aquatic Microbes* [I. Joint (ed.)]. Berlin, Springer-Verlag, pp. 278-302.

Kohut, J.T., S.M. Glenn, and R.J. Chant, 2004. Seasonal current variability on the New Jersey inner shelf. *Journal of Geophysical Research*, 109:C07S07, doi:10.1029/2003JC001963.

Lee, T.N., and L.J. Pietrafesa, 1987. Summer upwelling on the southeastern continental shelf of the U.S.A. during 1981. *Progress in Oceanography*, 19:276-312.

Lee, T., J. Yoder, and L. Atkinson, 1991. Gulf Stream frontal eddy influence on productivity of the southeast U.S. continental shelf. *Journal of Geophysical Research*, 96:22191-22205.

Li, N., J. Zhoa, P.V. Warren, J.T. Warden, D.A. Bryant, and J.H. Golbeck, 1991. PsaD is required for the stable binding of PsaC to the photosystem 1 core protein of Synechococcus sp. PCC 6301. *Biochemistry*, 30:7863-7872.

Linder, C., and G. Gawarkiewicz, 1998. A climatology of the shelfbreak front in the Middle Atlantic Bight. *Journal of Geophysical Research*, 103:18405-18423.

Linder, C., G. Gawarkiewicz, and R.S. Pickart, 2004. Seasonal characteristics of bottom boundary layer detachment at the shelfbreak front in the Middle Atlantic Bight. *Journal of Geophysical Research*, 109:C03049, doi:10.1029/2003JC002032.

Loder, J.W., B. Petrie, G. Gawarkiewicz, 1998. The coastal ocean off northeastern North America: A large scale view. In: *The SeOa*, Vol. 11. John Wiley & Sons, Inc., pp. 105-133.

Lohrenz, S.E., D.G. Redalje, P.G. Verity, C.N. Flagg, and K.V. Matulewski, 2002. Primary production on the continental shelf off Cape Hatteras, North Carolina. *Deep-Sea Research II*, 49:4479-4509.

Lozier, M.S., and G. Gawarkiewicz, 2001. Cross-frontal exchange as evidenced by surface drifter trajectories. *Journal of Physical Oceanography*, 31:2498-2510.

Malone, T.C., D.J. Conley, T.R. Fisher, P.M. Glibert, L.W. Harding, and K.G. Sellner, 1996. Scales of nutrient-limited phytoplankton productivity in Chesapeake Bay. *Estuaries*, 19:371-385.

Malone, T. C., P.G. Falkowski, T.S. Hopkins, G.T. Rowe, and T.E. Whitledge, 1983. Mesoscale response of diatom populations to a wind event in the plume of the Hudson River. *Deep-Sea Research*, 30:149-170.

Menzel, D.W., L.R. Pomeroy, et al. 1993. Introduction. In: *Ocean Processes: US Southeast Continental Shelf.* National Technical Information Service, Springfield, VA (DE93010744).

O'Reilly, J.E. and D.A. Busch, 1984. Phytoplankton primary production on the northwestern Atlantic shelf. *Rapports et Proces Verbaux des Reunions Conseil International pour l'Exploration de la Mer*, 183:255-268.

Reid, P.C., N.P. Holliday, and T.J. Smyth, 2001. Pulses in the eastern margin current and warmer water off the north west European shelf linked to North Sea ecosystem changes. *Marine Ecology Progress Series*, 215:283-287.

Ullman, D.S., and D. L. Codiga, 2004. Seasonal variation of a coastal jet in the Long Island Sound Outflow Region based on HF radar and Doppler current observations. *Journal of Geophysical Research*, 109:C07S06; doi:10.1029/2002JC001660.

Yoder, J.A., L.P. Atkinson, T.N. Lee, H.H. Kim, and C.R. McClain, 1981. Role of Gulf Stream frontal eddies in forming phytoplankton patches on the outer southeastern shelf. *Limnology and Oceanography*, 26:1103-1110.

Yoder, J.A., L.P. Atkinson, S.S. Bishop, J.O. Blanton, T.N. Lee, and L.J. Pietrafesa, 1985. Phytoplankton dynamics within Gulf Stream intrusions on the southeastern United States continental shelf during summer, 1981. *Continental Shelf Research*, 4:611-635.

Vandemark, D., J. Salisbury, C. Hunt, W. McGillis, J. Campbell, and F. Chai, 2006. The coastal Gulf of Maine as an atmospheric CO_2 sink with strong seasonal riverine control. *Eos*, 87(36), Ocean Sciences Meeting Supplement, Abstract OS35G-15.

Verity, P.G., D.G. Redalje, S.R. Lohrenz, C. Flagg, and R. Hristov, 2002. Coupling between primary production and pelagic consumption in temperate ocean margin pelagic ecosystems. *Deep-Sea Research II*, 49:4553-4569.

North America's Pacific Coast

Francisco P. Chavez
Monterey Bay Aquarium and Research Institute

Burke Hales
College of Oceanic and Atmospheric Sciences
Oregon State University

Martin Hernandez
Centro de Investigacion y de Educacion Superior de Ensenada

Niki Gruber
Institute of Biogeochemistry and Pollutant Dynamics
Department of Environmental Sciences
ETH Zürich

Debbie Ianson
Department of Fisheries and Oceans
Institute of Ocean Sciences

The west coast of the North American continent is a convergent margin with narrow, steep, largely rocky shelves. From Panama to the Gulf of Alaska, this coast covers the greatest latitudinal range of any of North America's coastal oceans (Figure 4.1). Weaker eastern boundary currents—the California Current in the south and the Alaska Current to the north—separate the open and coastal oceans. Here, wind forcing drives exchange with the open North Pacific Ocean. Wind-driven upwelling brings cool, high-pCO_2 and high-nutrient water from depth offshore to the coastal ocean surface, fueling strong productivity and biomass responses (Figure 4.2), and nutrient and CO_2-depleted waters are then returned offshore to the surface open ocean. Wind-driven downwelling forces waters with open-ocean character closer to the coastline. Eddies and meandering filaments of the wind-driven coastal currents and jets enhance exchange of coastal water with the open ocean, particularly in the mid-southern latitudes of the region (Figure 4.3). The effects of seasonal heating and cooling in this region are somewhat dampened by exchange with the open ocean, particularly the upwelling of cool waters during summer. River input to the west coast is important from northern California northward. This river discharge is highly seasonal, dominated by wintertime rainfall and springtime snowpack melt. Atmospheric circulation patterns result in mostly on-shore flow of air, although this is variable in the southern extent of the region.

Spatial and temporal biogeochemical variability is great on the west coast, and covers a range of scales. The presence of waters originating in the upper thermocline (depths as great as 200 m) of the North Pacific at depth on the shelf, and rapid productivity in surface waters, compress whole-ocean ranges of variability in parameters like pCO_2, O_2, and nutrient concentrations over the small vertical (tens of meters)

Figure 4.1. Map of the North American Pacific coastline, showing the great latitudinal extent and the narrow shelves.

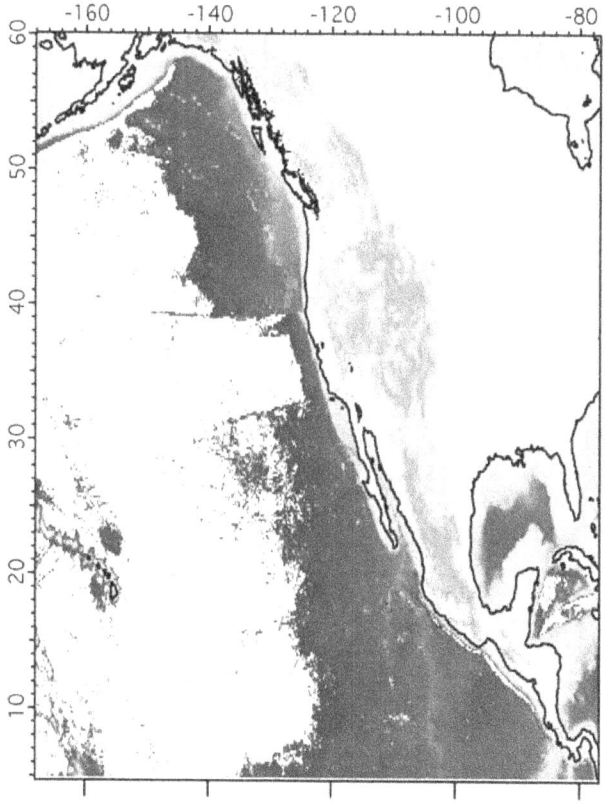

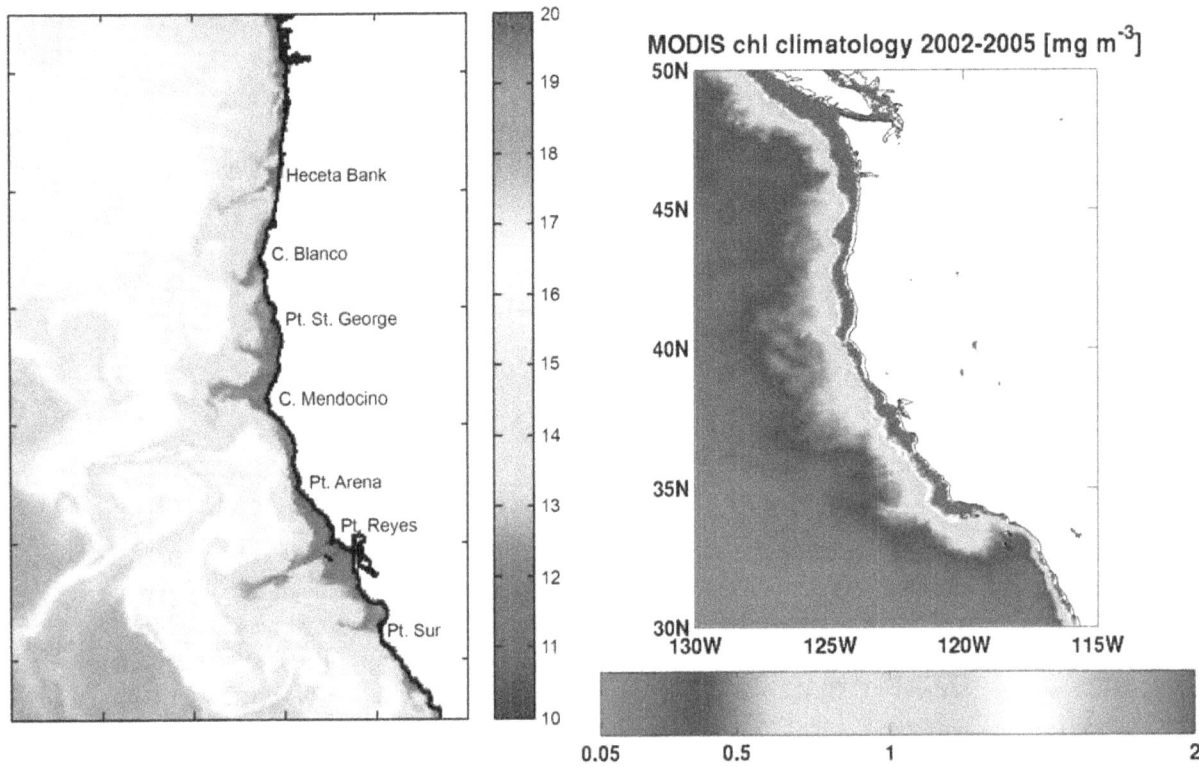

Figure 4.2. Satellite images of the US Pacific Coast, showing the effects of upwelling on left) surface-water temperatures, and right) chlorophyll abundance.

Figure 4.3. Satellite images of SST in the SCC, showing numerous eddy- and filament-like structures originating from the coastline.

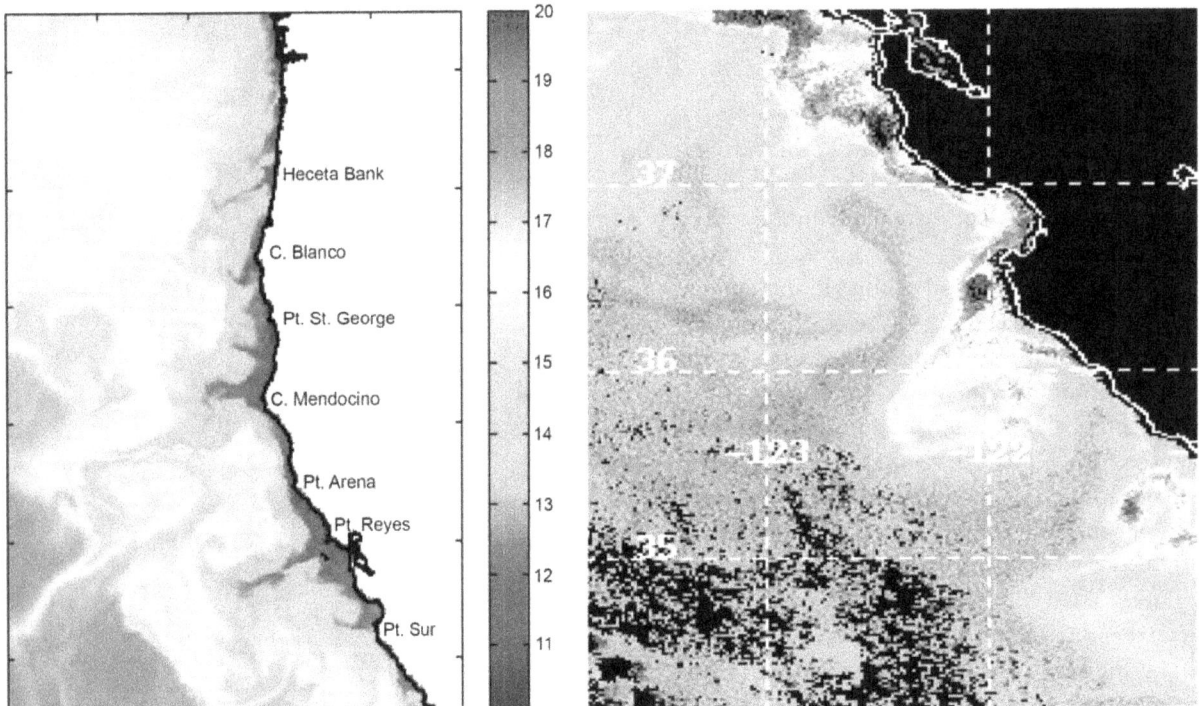

and cross-shelf (tens of kilometers) scale of the system (Friederich et al. 2002; Hales et al. 2005a, b, 2006; Figure 4.4). Near-shore pCO$_2$ levels can exceed 700 µatm, but can drop rapidly to values below 200 µatm within several kilometers seaward. The simple existence of either of these pCO$_2$ levels in the open ocean would be remarkable; their presence in such close proximity is probably unique to productive coastal settings. Alongshore variability can be short as well. In regions influenced by river plumes, alongshore variability scales can be as short as several kilometers. Another key scale of horizontal spatial variability is associated with the eddies that drive much of the interaction between the coastal and open ocean, ranging from 10 to 100 km. Temporal variability also covers a wide range of scales

from hourly (e.g., tidal) to interdecadal. An example of temporal variability is also given in Figure 4.4, where the nearshore pCO$_2$ off the Oregon coast drops from 700 to 200 µatm in the course of two days in response to a brief relaxation in upwelling conditions. Longer time-scale variability has been shown to impact the conditions of the west coast, ranging from interannual (e.g., Chavez et al. 2002; Friederich et al. 2002) to interdecadal (e.g., Peterson and Schwing, 2003; Huyer, 2003).

Subregions of North America's Pacific Coast

While alongshore spatial variability can be as short as several kilometers, significant differences exist

Figure 4.4. Cross-shelf transects of surface pCO$_2$ in the NCC and SCC, showing the high spatial and temporal variability.

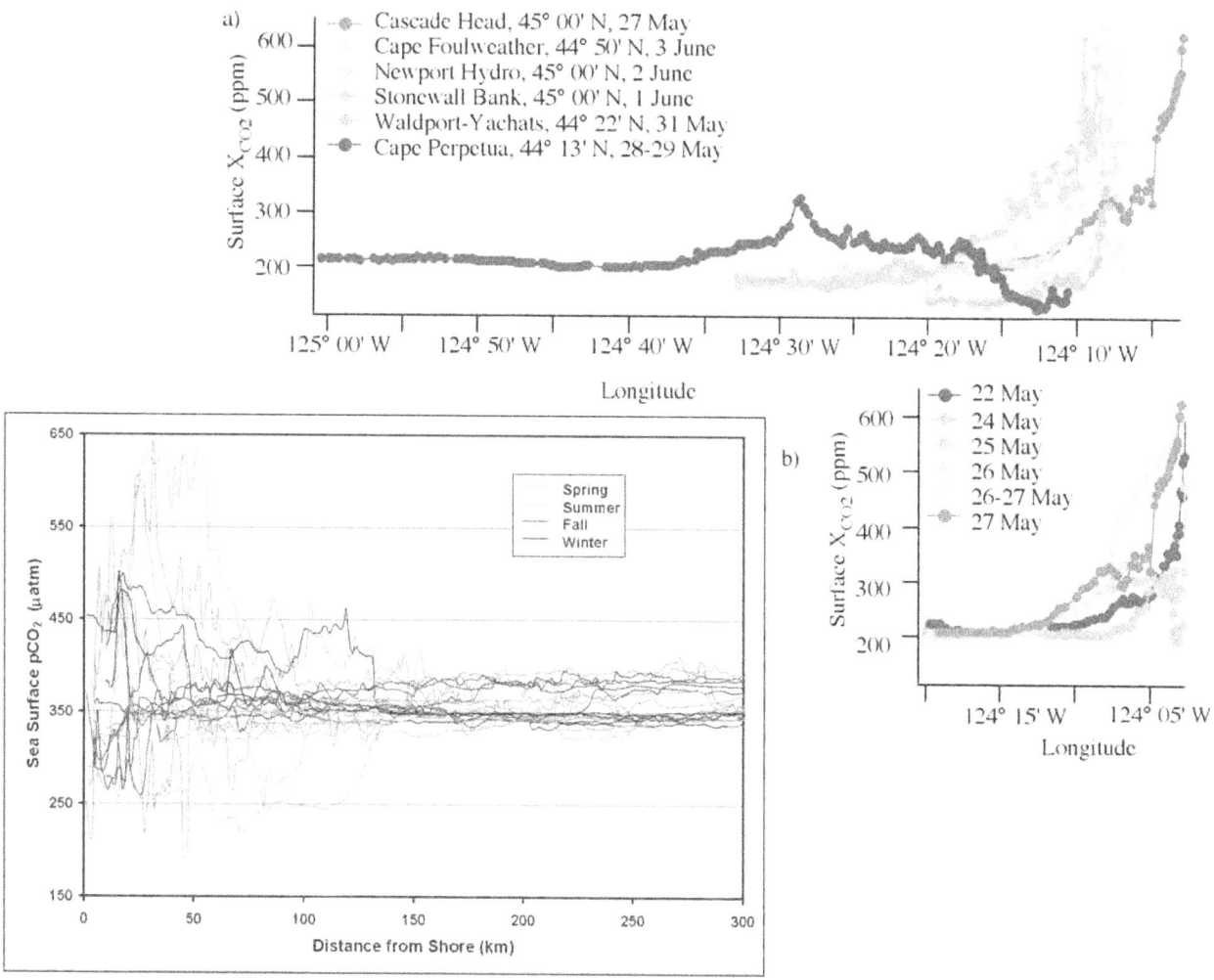

between the larger subregions of the Central American Isthmus (CAI), southern California Current (SCC), northern California Current (NCC), and the Gulf of Alaska (GoA).

The Pacific coast of the CAI, which comprises the eastern border of the tropical Pacific Ocean, between 5°N and 21°N is influenced by intense wind events that produce upwelling at three sites off Northern and Central America (Lavín et al. 1992): the Gulf of Tehuantepec in Mexico, the Gulf of Papagayo in the Costa Rica-Nicaragua border, and the Gulf of Panamá. These strong, focused winds can last from 1 to 15 days and blow offshore over the Pacific Ocean, generating filamentous ocean jets and high waves along their trajectory. Their effect favors strong near-shore oceanic mixing and intense lowering of sea-surface

temperatures (Trasviña et al. 1995). Another feature observed in this zone is the Costa Rica Dome (CRD), centered at 90°W and 9°N, which is a shoaling of the strong and shallow thermocline of the eastern Pacific Ocean. This phenomenon usually occurs in fall, winter, and early spring when polar air masses move south into the Gulf of Mexico and the Intertropical Convergence Zone (ITCZ) is at its southernmost position (Chelton et al. 2000; Figure 4.10; Gonzalez-Silvera et al. 2004).

In the Gulf of Tehuantepec, northern strong wind jets (15-20 m s⁻¹), called the "Tehuanos," are due to sea level pressure differences set up across the Isthmus of Tehuantepec by high-pressure systems that originate over the United States; the height of the Sierra Madre mountain range drops from 2000 to 250 m at the 40-km wide Chivela Pass, allowing the formation of

Figure 4.5. Depiction of offshore winds' effect on atmospheric CO_2 content at a mooring in the California Borderlands southeast of Los Angeles. In this figure, the arrow's directions represent wind direction, and their length represents the deviation of atmospheric CO_2 from the marine-air background (368 ppm). The largest positive deviations correspond to air sourced from the adjacent major metropolitan areas, and can exceed background pCO_2 by up to 60 ppm. After Leinweber et al. (submitted); A. Leinweber (pers. comm).

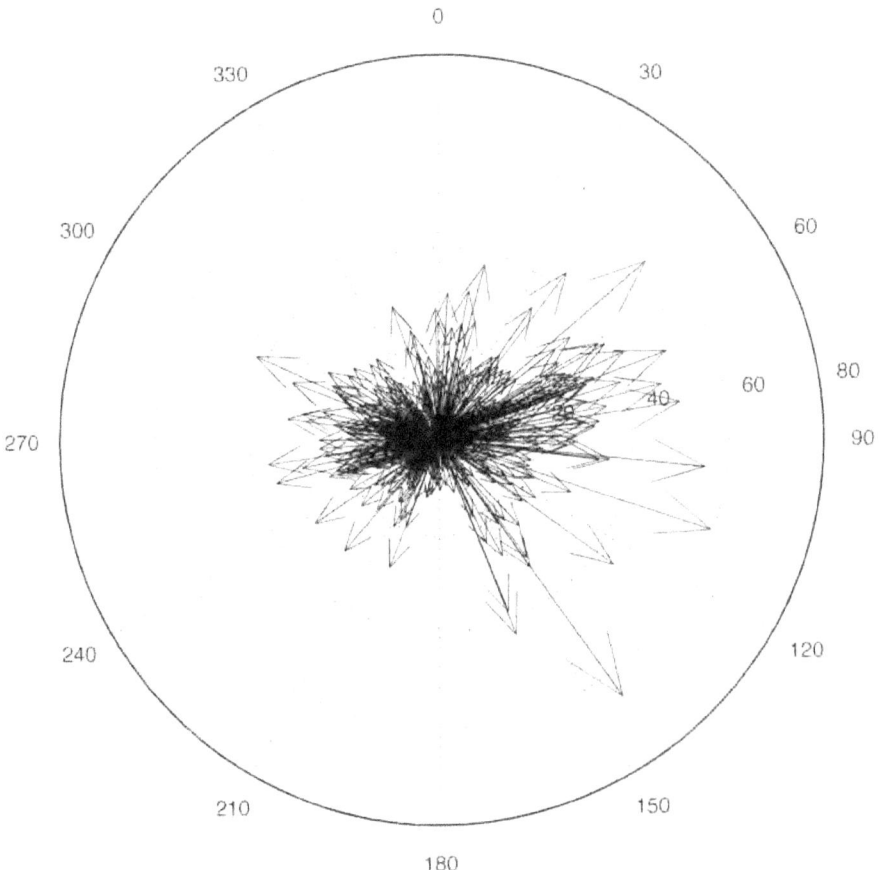

the wind jets. These winds have a maximum reach exceeding 500 km (Romero-Centeno et al. 2003). The jet shape of the wind causes an uneven Ekman transport, which accumulates surface water on the western side of the gulf, depressing the thermocline and producing anticyclonic and cyclonic eddies over the ocean with diameters ranging between 100 and 450 km (Figure 4.11; Gonzalez-Silvera et al. 2004). But also, under the wind jet, large plumes of cold water with high nutrients in the surface layer promote high chlorophyll concentration (Trasviña et al. 1995; Robles-Jarero and Lara-Lara, 1993).

In the SCC, which covers the entire coast from San Francisco Bay south to the southern tip of the Baja Peninsula, the shelf is quite narrow—barely a few kilometers across, e.g., at Monterey, California. The shelf in this region is so narrow that many of the signatures of coastal processes, like elevated chlorophyll and filaments of cool upwelled waters, exist over water columns seaward of the bathymetric shelf break. Riverine input to the SCC is limited, and thermal heating is stronger than in higher latitude regions. Circulation in the SCC is dominated by upwelling-favorable, equatorward wind-forcing conditions year-round (e.g., Ware and Thompson, 2005). Eddy formation is significant (Figure 4.3), here, more so than in the NCC. Winds periodically blow offshore in the southern and Baja California region, which can have a significant effect on atmospheric composition over the coastal ocean (Figure 4.5; Leinweber et al. submitted). Anthropogenic activity is pronounced in the central-northern part of this region, including the metropolitan areas of San Francisco, Los Angeles, and San Diego.

The NCC, stretching from Northern California to Vancouver Island has significantly broader shelves than the CAI or SCC, ranging up to 100 km wide off the Washington coast. The northern NCC also marks the onset of extensive island- and peninsula-sheltered coastal waters, including the Strait of Juan de Fuca, Puget Sound, and the inside passage shoreward of Vancouver Island. Tidal energies in these inside-passage waters are strong, and the resulting mixing is intense. Winds are almost always on- and along-shore in either upwelling or downwelling conditions, with few events that transport continental air over the coastal ocean.

Wind-driven circulation is strongly seasonal—upwelling dominates summer months, while downwelling dominates winter (e.g., Barth and

Wheeler, 2005). These two circulation regimes result in strong alongshore transport in the form of coastal jets that set themselves up between the mid-shelf and the shelf break (Allen and Newberger, 1996). During summer, upwelling appears to deliver deep-source waters far inshore of the equatorward-flowing coastal jet (Perlin et al. 2005), and the coldest, saltiest, most nutrient-rich surface waters are found in the near-shore region (van Geen et al. 2000; Hales et al. 2005b). During winter conditions, it appears that most of the downwelling circulation occurs seaward of the jet, and large portions of the inner shelf are essentially isolated from exchange with the adjacent open ocean (Allen and Newberger, 1996; Austin and Barth, 2002; Barth, 2003). Eddy-driven circulation is not as strong here as in the SCC. This may be due to the broader shelves that keep large-scale ocean circulation at the shelf break further from shore as a result of conservation of potential vorticity (e.g., Gill, 1982), or simply due to stronger winds in the SCC (e.g., Ware and Thompson, 2005).

Figure 4.6. Depiction of the dramatic increase in river discharge to the coastal ocean north of San Francisco Bay, and the correspondence to coastal chlorophyll abundance. Circle size scales to river discharge at seaward-most USGS gauging stations. From Chase et al. 2007.

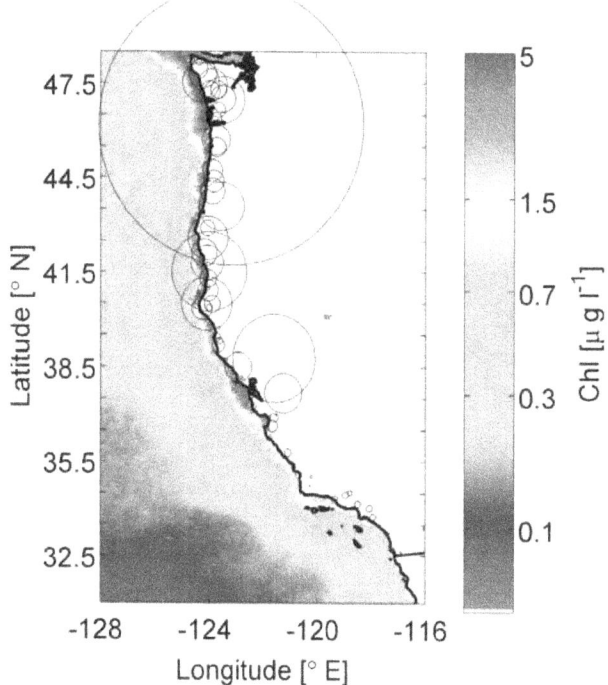

The region receives large riverine inputs (Figure 4.6; Chase et al. 2007) of naturally high-nutrient water, the result of nitrogen fixation in the forested uplands (e.g., Compton et al. 2003). Some of these are from numerous small coastal rivers that deliver most of their discharge in winter when heavy rain falls on low-altitude coastal mountain ranges, and it appears that these inputs are only weakly modified by estuarine processes (Chase et al. 2007; Wetz et al. 2006). Other rivers (e.g., the Sacramento and Columbia) drain higher, colder, inland mountain ranges and have greater snow-melt forcing in mid-late spring. Still others, notably at the northern end of this region, drain into the secluded waters of Puget Sound and the Inside Passage inshore of Vancouver Island, and their integrated effects on the open coastal ocean are seen in passages such as the Strait of Juan de Fuca and the Strait of Georgia.

There is less direct anthropogenic impact on the coastal ocean of the NCC, with the smaller (relative to the large populations of Southern California)

metropolitan areas of Portland, Seattle, and Vancouver physically removed from the open coastline by hundreds of kilometers; however the larger river systems (especially the Columbia) have been extensively dammed, significantly altering their water, sediment, and nutrient discharge patterns (e.g., Ebbesmeyer and Tangborn, 1992; Sullivan et al. 2001).

The Gulf of Alaska subregion begins north of Vancouver Island and extends north and west along the Canadian and Alaskan coastline to the Aleutian Archipelago. Its northern boundary reaches within several hundred kilometers of the Arctic Circle, and its equivocal western boundary encompasses the international date line. This region consists of over 5000 km of simple coastline, and contains the widest shelves in North America's Pacific margin, reaching over 300 km width south of Prince William Sound. As a result, this area contains the majority of the shelf area in the Pacific margin. The coastline is rocky and complex, containing innumerable fjords and small islands, in addition to several large islands, inlets, and sounds that contribute to an extensive 'inland waterway' that is sheltered to some extent from the open Pacific Ocean. This area is strongly impacted seasonally, both by strong winter storms and by the extreme variation in insolation. Many rivers, fed by snowpack and glacier melt, drain into the Gulf of Alaska and this, combined with significant net precipitation delivered directly to the sea surface, significantly impacts the stratification and circulation of the region. This area is strongly impacted by shifts in the PDO (Figure 4.7) and is likely to experience significant changes resulting from a warming climate, as high-latitude regions are predicted to experience disproportionate changes in heating, precipitation, and river discharge.

Figure 4.7. Map of SST deviation before and after the PDO regime shift in 1977. Note the disproportionate impact on GoA waters, even in summer. From http://globec.oce.orst.edu/nepsummary.html, courtesy of Franklin Schwing.

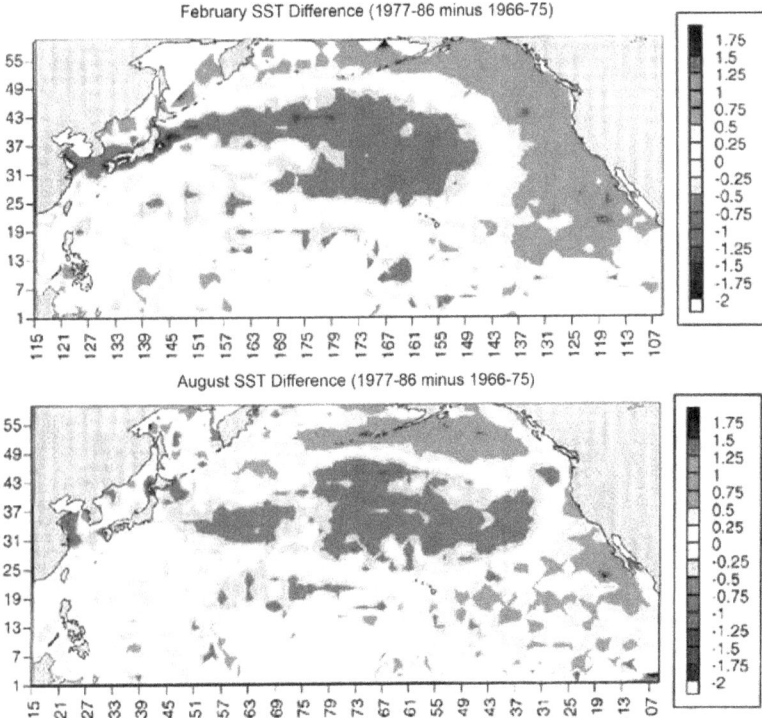

Circulation here is mostly convergent, dominated by currents that originate with the eastward flowing West Wind Drift and Subarctic Current in the open North Pacific. At approximately the latitude of the north tip of Vancouver Island, this eastward

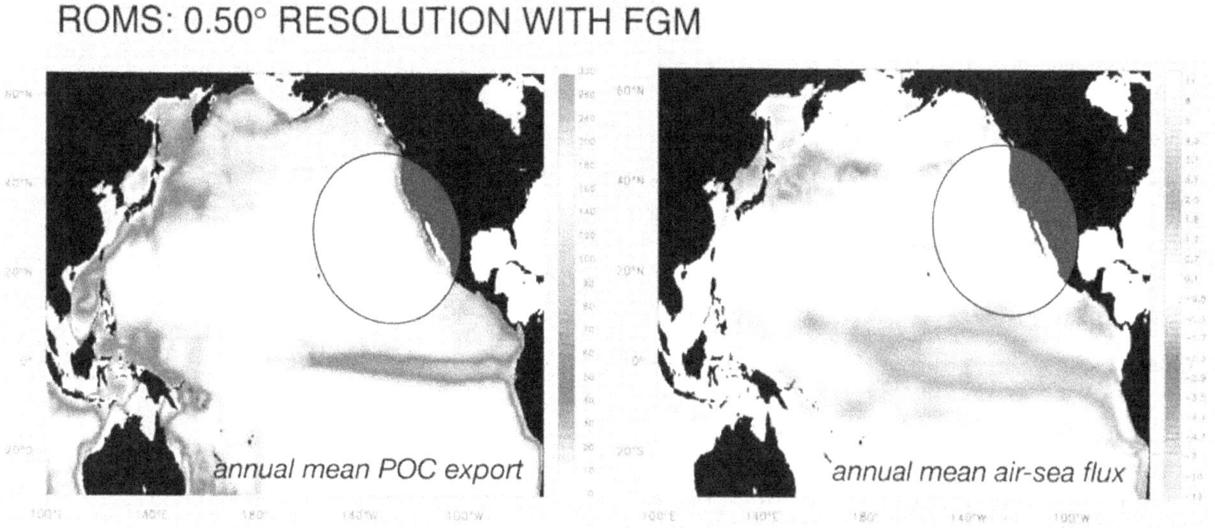

ROMS: 0.50° RESOLUTION WITH FGM

annual mean POC export

annual mean air-sea flux

Figure 4.8. Model results showing a) air-sea uptake of CO_2 and b) deep-ocean carbon export off the Pacific coast of North America. Note that while the air-sea uptake of CO_2 is equivocal even as to its sign, the carbon export is significantly enhanced over the extent of this region. Figure courtesy N. Gruber.

flow bifurcates as it approaches the North American continent to form the southward-flowing California Current and the northward-flowing Alaska Current. The latter flows north and west along the Canadian and Alaskan coastlines until it begins to flow predominantly westward at Prince William Sound. There, with contributions from the circulation of the Alaska Gyre, this current becomes the west-southwest-ward-flowing Alaskan Stream (Thompson and Ware, 1989). Baroclinic instabilities in this region can drive large eddy features that enhance exchange with the open ocean (e.g., Di Lorenzo et al. 2004). Tidal transport is an important mechanism here, especially in the exchange of fjord and inside passage waters with the open coastal ocean.

Carbon Cycling on the Pacific Coast

The high nutrient content of waters from the North Pacific drives huge primary productivity responses along the west coast. Net to new productivity ratios for coastal diatoms seems to be high (Dugdale et al. 1990, 2006) but also highly variable (Kokkinakis and Wheeler, 1987), so export productivity is probably high as well (Hales et al. 2006; Gruber et al. 2006; Figures 4.8 and 4.9). In the NCC, upwelled

macronutrients appear to be completely consumed (Chase et al. 2005; Hales et al. 2005b); however in the SCC, HNLC conditions sometimes exist, presumably the result of iron limitation (Hutchins et al. 1998; Dugdale et al. 2006). Sediment burial and diagenesis of carbon is low relative to export productivity here (Hartnett and Devol 2003; Hales et al. 2006). Wintertime input of terrestrial OC may be significant in winter in the NCC (Wetz et al. 2006; Chase et al. 2007). A detailed data-driven budget of O_2 and POC off the Oregon coast suggests that a large fraction of the summertime upwelling-fueled productivity is exported off-shelf to the adjacent deep ocean (Table 4.1; Figure 4.9). Modeling results off central California in the SCC (Figure 4.12; Gruber et al. 2006) show similar export fluxes.

Net Air-Sea CO_2 Exchange

The net effect of these processes on the exchange of CO_2 with the atmosphere is complicated and spatially and temporally variable. The high-CO_2 and high-nutrient content of upwelled waters allows for both the possibility of strong CO_2 efflux to the atmosphere and strong uptake by the surface ocean as coastal phytoplankton consume the abundant nutrients.

Table 4.1. An organic carbon budget for the central Oregon coast during upwelling season. Adapted from Hales et al. 2006.

Budget term	Value (mmol C m^{-2} d^{-1})	Uncertainty (mmol C m^{-2} d^{-1})
Cross-shelf transport	-3	-0.3 to 3
Along-shore transport	-27	-100 to 27
Temporal change	-5	-10 to 5
Sediment consumption + burial	-15	±15
Net production	140	-90 to 120
Offshore export	-100	-70 to 40

Figure 4.9. Data-based budget for C export from the NCC from Hales et al. (2006). O$_2$ production suggests large net production over the summer, but POC budgets do not account for an equivalent increase in standing stock. Implied export is over ten times greater than the atmospheric uptake of CO$_2$ given by Hales et al. (2005a).

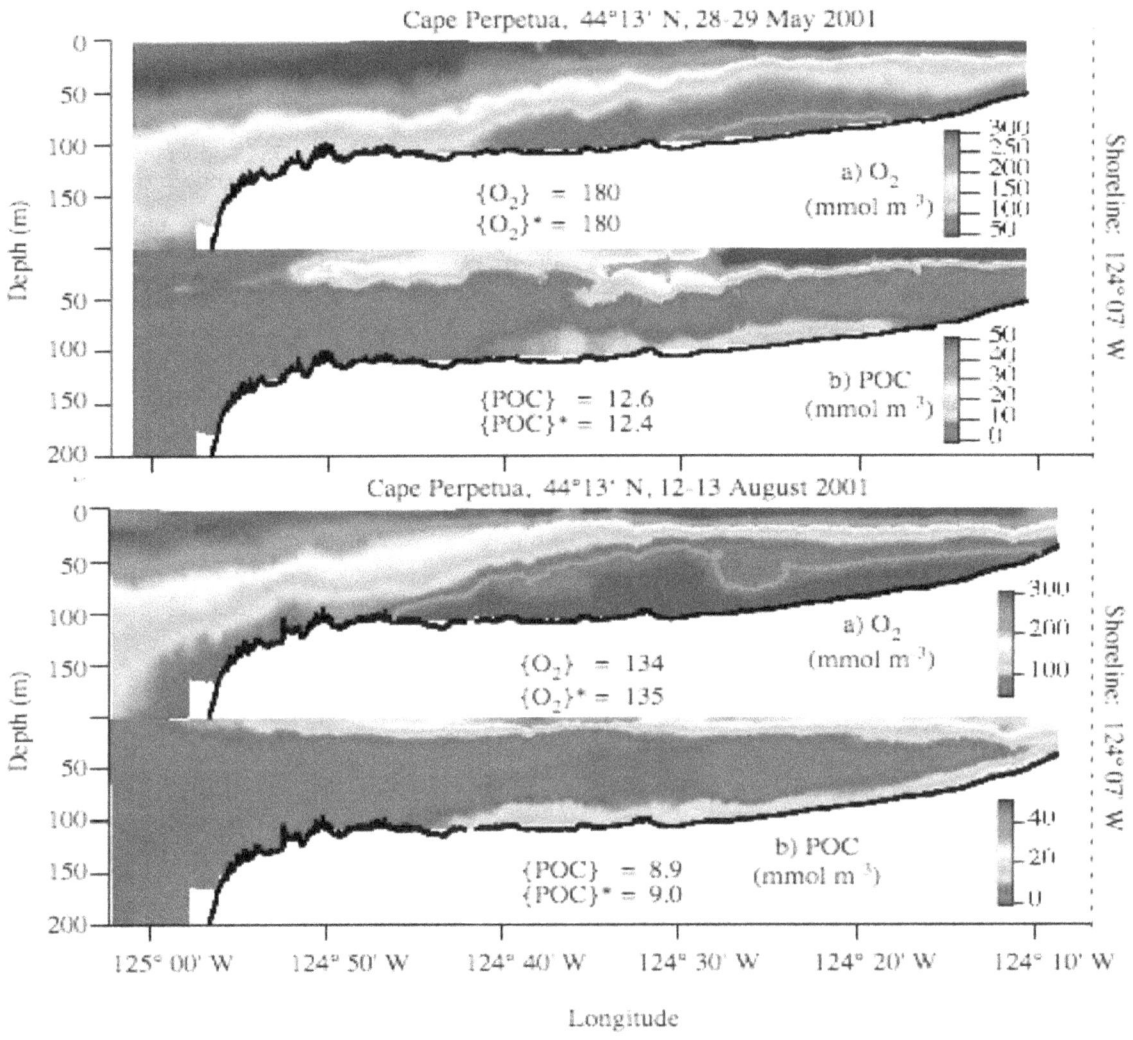

In modeled representations of the carbon cycle, net air-sea CO_2 exchange is weak (Figure 4.9), despite the large modeled export productivity. The net exchange seems to have strong latitudinal dependence on this coast as well as on the Atlantic Coast: in the Southern California Current system, coastal waters appear to be a near-zero or slightly positive annually-averaged source of CO_2 to the atmosphere (Leinweber et al. submitted; Friederich et al. 2002, 2006); to the north, productivity appears to be high enough to consume regenerated and preformed nutrients alike, drawing CO_2 down to levels far below atmospheric saturation (Friederich et al. 2006; Hales et al. 2005a; Ianson et al. 2003).

Research Programs in Pacific Coastal Waters

The Pacific Coast has been the subject of some long-running interdisciplinary research programs. The most exemplary of these is the California Cooperative Oceanic Fisheries Investigations (CALCOFI) program (http://www.calcofi.org), a cooperative partnership between the California Department of Fish and Game, the NOAA Fisheries Service, and the Scripps Institution of Oceanography. This program has conducted regular cruises since 1949 to investigate the chemical, physical, and biological conditions off the California coast between Point Conception and San Diego.

Recently, the US Global Ocean Ecosystems Dynamics (GLOBEC; http://www.usglobec.org) program, a cooperative effort between the NSF, the NOAA Center for Sponsored Coastal Ocean Research, and the NOAA National Marine Fisheries Service, sponsored an extensive research effort in the northeast Pacific (NEP); (http://globec.oce.orst.edu/nepsummary.html). The GLOBEC NEP program included study of both the California Current System (CCS) and the Coastal Gulf of Alaska (CGOA) with eight-year programs, including synthesis and modeling components, that began for the CCS in 1998 and the CGOA in 2000. The goal of this ongoing effort is to gain understanding of the effects of climate variability on these ecosystems and to develop predictive capabilities that will allow predictions of their responses to future climate change. This has been pursued through a combination of monitoring and process studies, and the development of predictive models.

Figure 4.10. Monthly composites of chlorophyll-a concentration (mg m⁻³) from SeaWiFS and sea-surface temperature (SST, 1°C) from AVHRR. The mean position of the Costa Rica Dome (from Fiedler, 2002) is superimposed. From Gonzalez-Silvera et al.. (2004).

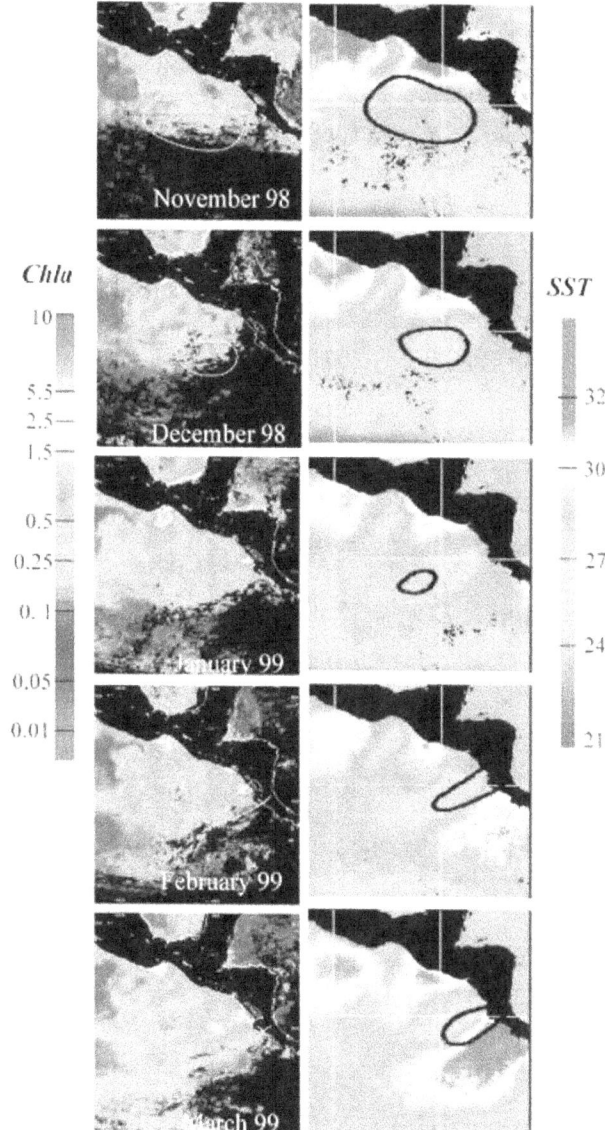

More recently still, the Coastal Ocean Processes (CoOP; http://www.skio.peachnet.edu/coop/) program within the NSF supported three focused studies in the NCC subregion. Two of these, the Coastal Ocean Advances in Shelf Transport (COAST; Barth and Wheeler, 2005; http://damp.coas.oregonstate.edu/coast/) and Wind Events and Shelf Transport (WEST; Largier et al. 2006; http://www.skio.usg.edu:443/research/coop/current.php) studied the wind-driven circulation, particularly upwelling, off the coasts of Oregon and Northern California, respectively, and the effects of that circulation on biological productivity. These were five-year projects beginning in 2000. The

third, the River Influences on Shelf Ecosystems (RISE; http://www.ocean.washington.edu/rise) is an ongoing five-year project that began in 2003. Another new project focusing on the Columbia River and nearby coastal ocean has been funded through the NSF Science and Technology Centers program. This program, (CMO; http://www.stccmop.org) is a five-year effort. Finally, the North American Pacific coast is slated to be a major part of the Ocean Observatories Initiative (OOI), according to the Ocean Research Interactive Observatory Networks (ORION) Program report. Proposed work includes extensive implementation of a variety of observational strategies (Figure 4.13).

Figure 4.11. SeaWiFS LAC images of chlorophyll-a concentration (mg m⁻¹) showing identified eddies. Eddies were labeled according to type (anticyclonic, A, or cyclonic, C), origin (Tehuantepec, T, or Papagayo, P) and date of generation. From Gonzalez-Silvera et al.. (2004).

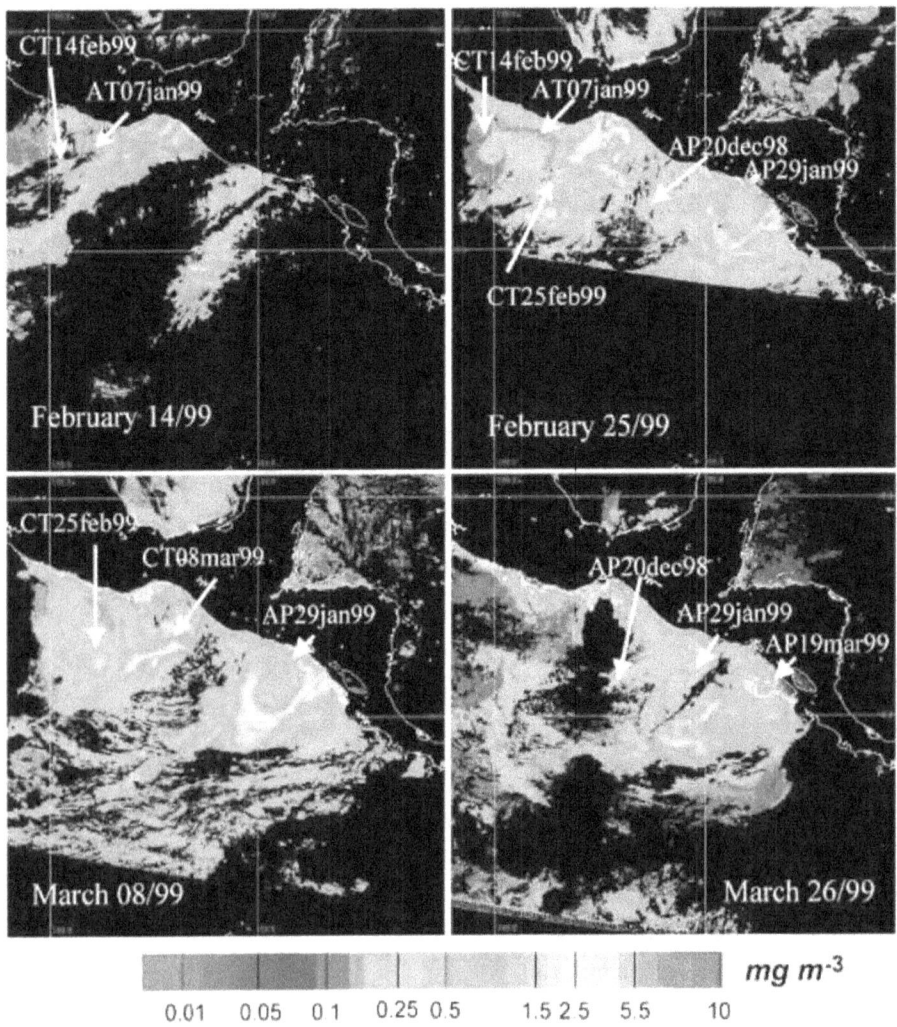

Figure 4.12. Model-based budget for the SCC. Note the magnitude of export fluxes relative to air-sea exchange.

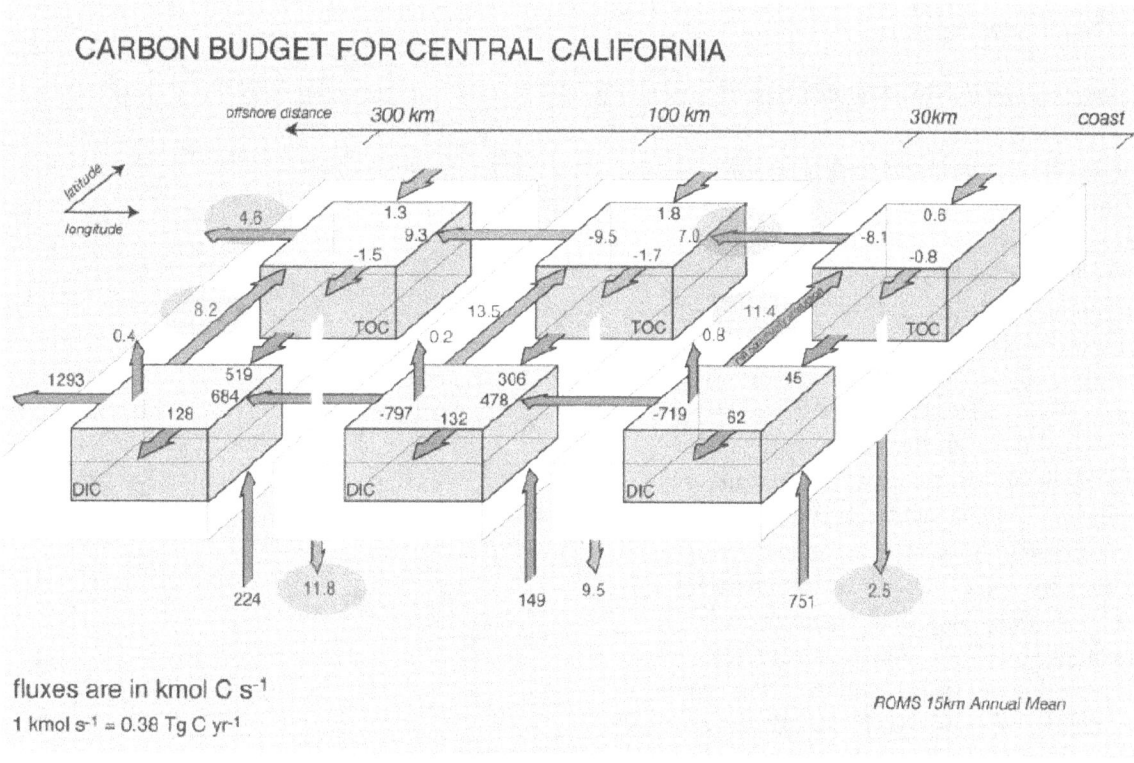

Figure 4.13. Schematic of proposed observatory systems on the US west coast, and their relation to historical programs and existing resources.

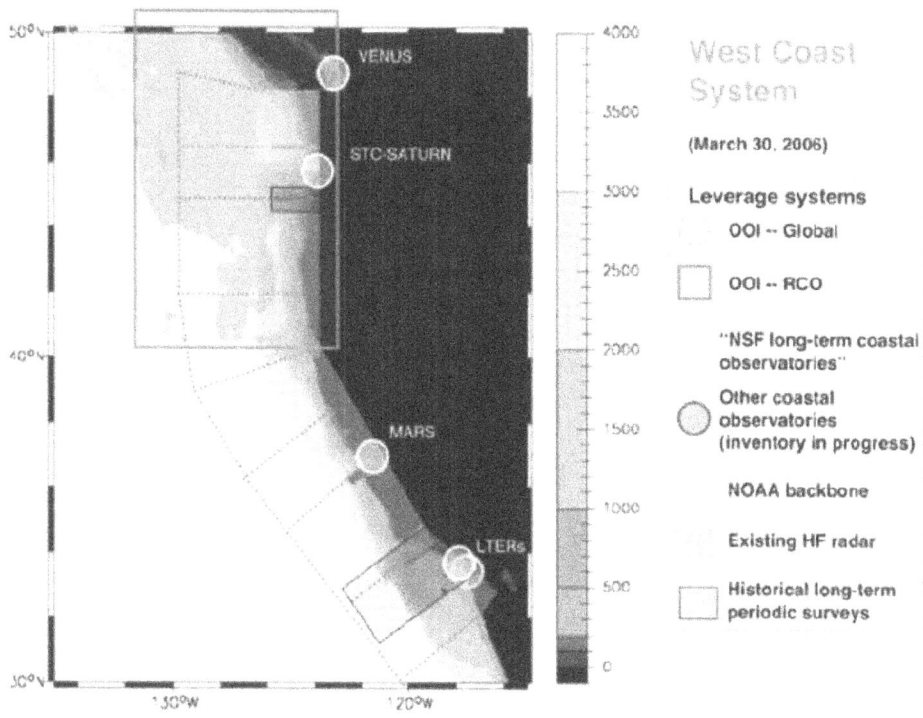

Of these historical, ongoing, and proposed efforts, only the COAST and CMOP projects have included inorganic carbon measurements.

Anthropogenic Impacts on the Pacific Coast

Despite the close connections to the vast open Pacific Ocean, this coast does have some sensitivities to anthropogenic activities. Numerous activities—from harvesting sea otters (e.g., Estes and Palmisano, 1974) to damming rivers (e.g., Sullivan et al. 2001; Ebbesmeyer and Tangborn, 1992) to discharging urban waste (e.g., Lee and Wiberg, 2002)—can have direct impacts on the greater Pacific Coast ecosystem that extend beyond the immediate perturbation. The effects of global climate change may be significant as well. Warming and freshening of high-latitude regions where upwelled density horizons outcrop may change 'preformed' nutrient and dissolved gas concentrations, while increased warming of upwelled waters could impact gas solubility. There is some indication that higher winds could be associated with the greater land-sea thermal gradient of a warmed climate (Zwiers et al. 2002). River discharge to the coastal ocean could be significantly affected, particularly at higher latitudes where climate models predict increased precipitation (Arora and Boer, 2001).

References

Allen, J.S., and P.A. Newberger, 1996. Downwelling circulation on the Oregon continental shelf. Part I: Response to idealized forcing. *Journal of Physical Oceanography*, 26:2011-2035.

Arora, V.K., and G.J. Boer, 2001. Effects of simulated climate change on the hydrology of major river basins. *Journal of Geophysical Research*, 106(D4):3335-3348.

Austin, J.A., and J.A. Barth, 2002. Drifter behavior on the Oregon-Washington shelf during downwelling-favorable winds. *Journal of Physical Oceanography*, 32:3132-3144.

Barth, J.A., and P. A. Wheeler, 2005. Introduction to special section: Coastal Advances in Shelf Transport. *Journal of Geophysical Research*, 110:C10S01, doi:10.1029/2005JC003124.

Barth, J.A., 2003. Anomalous southward advection during 2002 in the northern California Current: Evidence from Lagrangian surface drifters. *Geophysical Research Letters*, 30:8024, doi:10.1029/2003GL017511.

Chase, Z., B. Hales, T.J. Cowles, R. Schwartz, and A. van Green, 2005. Distribution and variability of iron input to Oregon coastal waters during the upwelling season. *Journal of Geophysical Research*, 110:C10S12, doi:10.1029/2004JC002590.

Chase, Z., P. Strutton, and B. Hales, 2007. Iron links river runoff and shelf width to phytoplankton biomass along the US west coast. *Geophysical Research Letters*, 34:L04607, doi:10.1029/2006GL028069.

Chavez, F.P., C.A. Collins, A. Huyer, and D.I. Mackas, 2002. El Niño along the west coast of North America. *Progress in Oceanography*, 54:1-5.

Chelton, D.B., M.H. Freilich, and S.K. Esbensen, 2000. Satellite observations of the wind jets off Central America. Part II: regional relationships and dynamical considerations. *Monthly Weather Review*, 128:2019-2043.

Compton, J.E., M.R. Church, S.T. Larned, and W.E. Hogsett, 2003. Nitrogen export from forested watersheds in the Oregon Coast Range: the role of N_2-fixing red alder. *Ecosystems*, 6:773-785.

Daly, K., R. Jahnke, M. Moline, R. Detrick, D. Luther, G. Matsumoto, L. Mayer, and K. Raybould, 2006. *Report of the Design and Implementation Workshop*, 27-30 March 2006. http://www.orionprogram.org/PDFs/DI_report_final.pdf.

Di Lorenzo, E., A.J. Miller, D.J. Neilson, B.D. Cornuelle, and J.R. Moisan, 2004. Modeling observed California Current mesoscale eddies and the ecosystem response. *International Journal of Remote Sensing*, 25:1307-1312.

Dugdale, R.C., F.P. Wilkerson, and A. Morel, 1990. Realization of new production in coastal upwelling areas: A means to compare relative performance. *Limnology and Oceanography*, 35:822-829.

Dugdale, R.C., F.P. Wilkerson, A. Marchi, and V.E. Hogue, 2006. Nutrient controls on new production in the Bodega Bay, California, coastal upwelling plume. *Deep Sea Research II*, 53:3049-3062.

Ebbesmeyer, C.C., and W. Tangborn, 1992. Linkage of reservoir, coast, and strait dynamics, 1936-1990: Columbia River basin, Washington coast, and Juan de Fuca Strait. In: *Interdisciplinary Approaches in Hydrology and Hydrogeology*. American Institute of Hydrology, pp. 288-299.

Estes, J.A., and J.F. Palmisano, 1974. Sea otters: Their role in structuring nearshore communities. *Science*, 185:1058-1060.

Fiedler, P.C., 2002. The annual cycle and biological effects of the Costa Rica Dome. *Deep-Sea Research I*, 49:321-338.

Friederich, G.E., F.P. Chavez, P.M. Walz, and M.G. Burczynski, 2002. Inorganic carbon in the central California upwelling system during the 1997-1999 El Niño-La Niña event. *Progress in Oceanography*, 54:185-203.

Friederich, G.E., F.P. Chavez, G. Gaxiola, J. Ledesma, and O. Ulloa, 2006. Air-sea CO_2 fluxes in coastal upwelling regions along the west coasts of the Americas. In: *Proceedings of the 2006 ASLO Meeting*, 20-24 2006, Honolulu, HI.

Gill, A.E., 1982. *Atmosphere-Ocean Dynamics*. Academic Press, Inc., Orlando, Florida, 662 pp.

Gonzalez-Silvera, A., E. Santamaria-del-Angel, R. Millán-Nuñez, H. Manzo-Monroy, 2004. Satellite observations of mesoscale eddies in the Gulfs of Tehuantepec and Papagayo (Eastern Tropical Pacific). *Deep-Sea Research II*, 51:587-600.

Grantham, B.A., F. Chan, K.J. Nielsen, D.S. Fox, J.A. Barth, A. Huyer, J. Lubchenco, and B.A. Menge, 2004. Upwelling-driven nearshore hypoxia signals ecosystem and oceanographic changes in the northeast Pacific. *Nature*, 429:749-754.

Gruber, N., H. Frenzel, S.C. Doney, P. Marchesiello, J.C. McWilliams, J.R. Moisan, J.J. Oram, G.-K. Plattner, and K.D. Stolzenbach, 2006. Eddy-resolving simulation of plankton ecosystem dynamics in the California Current System. *Deep Sea Research Part I*, 53:1483-1516.

Hales, B., T. Takahashi, and L. Bandstra, 2005a. Atmospheric CO_2 uptake by a coastal upwelling system *Global Biogeochemical Cycles*, 19:GB1009, doi:10.1029/2004GB002295.

Hales, B., J.N. Moum, P. Covert, and A. Perlin, 2005b. Irreversible nitrate fluxes due to turbulent mixing in a coastal upwelling system. *Journal of Geophysical Research*, 110:C10S11, doi:10.1029/2004JC002685.

Hales, B., L. Karp-Boss, A. Perlin, and P. Wheeler, 2006. Oxygen production and carbon sequestration in an upwelling coastal margin. *Global Biogeochemical Cycles*, 20:GB3001, doi:10.1029/2005GB002517.

Hartnett, H.E., and A. Devol, 2003. Role of a strong oxygen-deficient zone in the preservation and degradation of organic matter: A carbon budget for the continental margins of northwest Mexico and Washington State. *Geochimica et Cosmochimica Acta*, 67:247-264.

Hutchins, D.A., G.R. DiTullio, Y. Zhang, and K.W. Bruland, 1998. An iron limitation mosaic in the California upwelling regime. *Limnology and Oceanography*, 43:1037-1054.

Huyer, A., 2003: Preface to special section on enhanced subarctic influence in the California Current, 2002. *Geophysical Research Letters*, 30:8019, doi:10.1029/2003GL017724.

Ianson, D., S.E. Allen, S.L. Harris, K.J. Orians, D.E. Varela, and C.S. Wong, 2003. The inorganic carbon system in the coastal upwelling region west of Vancouver Island, Canada. *Deep Sea Research I*, 50:1023-1042.

Kokkinakis, S.A., and P.A. Wheeler, 1987. Nitrogen uptake and phytoplankton growth in coastal upwelling regions. *Limnology and Oceanography*, 32:1112-1123.

Largier, J.L., C.A. Lawrence, M. Roughan, D.M. Kaplan, E.P. Dever, C.E. Dorman, R.M. Kudela, S.M. Bollens, F.P. Wilkerson, R.C. Dugdale, L.W. Botsford, N. Garfield, B. Kuebel Cervantes, and D. Koracin, 2006. WEST: a northern California study of the role of wind-driven transport in the productivity of coastal plankton communities. *Deep Sea Research II*, 53:2833-2849, doi:10.1016/j.dsr2.2006.08.018.

Lavín, M.F., J.M. Robles, M.L. Argote, E.D. Barton, R. Smith, J. Brown, M. Kosro, A. Trasviña, H.S. Vélez, and J. García, 1992. Fı´sica del Golfo de Tehuantepec. *Ciencia y Desarrollo*, 97-107.

Lee, H.J., and P.L. Wiberg, 2002. Character, fate, and biological effects of contaminated, effluent-affected sediment on the Palos Verdes margin, southern California: an overview. *Continental Shelf Research*, 22:835-840.

Leinweber, A., N. Gruber, H. Frenzel, G.E. Friederich, and F.P. Chavez, 2006. The impact on the estimation of the air-sea CO_2 flux in coastal areas. *Geophysical Research Letters*, submitted.

Perlin, A., J.N. Moum, and J. Klymak, 2005. Response of the bottom boundary layer over a sloping shelf to variations in alongshore wind. *Journal of Geophysical Research*, 110:C10S09, doi:10.1029/2004JC002500.

Peterson, W.T., and F.B. Schwing, 2003. A new climate regime in northeast pacific ecosystems. *Geophysical Research Letters*, 30:1896, doi:10.1029/2003GL017528.

Robles-Jarero, E.G., and J.R. Lara-Lara, 1993. Phytoplankton biomass and primary productivity by size classes in the Gulf of Tehuantepec, Mexico. *Journal of Plankton Research*, 15:1341-1358.

Romero-Centeno, R., J. Zavala-Hidalgo, A. Gallegos, and J.J. O'Brien, 2003. Isthmus of Tehuantepec wind climatology and ENSO signal. *Journal of Climate*, 16:2628-2639.

Sullivan, B.E., F.G. Prahl, L.F. Small, and P.A. Covert, 2001. Seasonality of phytoplankton production in the Columbia River: A natural or anthropogenic pattern? *Geochimica et Cosmochimica Acta*, 65:1125-1139

Thomson, R.E., and D.M. Ware, 1989. Oceanic factors affecting the distribution and recruitment of west coast fisheries. *Canadian Technical Report of Fisheries and Aquatic Sciences*, 1626:31-64.

Trasviña, A., E.D. Barton, J. Brown, H.S. Vélez, M. Kosro, and R.L. Smith, 1995. Offshore wind forcing in the Gulf of Tehuantepec, Mexico: the asymmetric circulation. *Journal of Geophysical Research*, 100:20649-20663.

van Geen, A., R.K. Takesue, J. Goddard, T. Takahashi, J.A. Barth, and R.L. Smith, 2000. Carbon and nutrient dynamics during coastal upwelling off Cape Blanco, Oregon. *Deep Sea Research II*, 47:975-1002.

Ware, D.M., and R.E. Thompson, 2005. Bottom-up ecosystem trophic dynamics determine fish production in the Northeast Pacific. *Science*, 308:1280-1284.

Wetz, M.S., B. Hales, Z. Chase, P.A. Wheeler, and M.M. Whitney, 2006. Riverine input of macronutrients, iron, and organic matter to the coastal ocean off Oregon, U.S.A., during the winter. *Limnology and Oceanography*, 51:2221-2231.

Zwiers, F.W., 2002. *The 20-year forecast*. Nature, 416:690-691.

North America's Gulf of Mexico Coast

Steve Lohrenz
University of Southern Mississippi

Wei-Jun Cai
University of Georgia

The Gulf of Mexico (GoM) is an enclosed subtropical marginal sea located at the southeastern corner of North America and is bordered by the United States to the north, Mexico to the west and south, and Cuba to the southeast (Figure 5.1). It has a total surface area of 1.5×10^6 km^2 with 80% of the area being continental margin (38% shallow water or <20 m; 22% mid- and outer continental shelf or 20-180 m, and 20% continental slope or 180-3000 m; Gore, 1992). The Gulf Coast's thermal and episodic-storm forcing is similar to the southern East Coast that is bordered by the Gulf Stream, but is distinct in many other ways. Its interaction with the open North Atlantic is due primarily to the Loop Current, which is the precursor of the Gulf Stream, flowing north and west along the west Florida shelf producing abundant mesoscale eddies and resulting in some upwelling along the shelf. Large-scale atmospheric circulation patterns show significant flow of air from the Gulf of Mexico northward onto the continent. The Gulf Coast is strongly influenced by riverine inputs. Several major rivers deliver nearly 1000 km^3 of freshwater per year into the Gulf (Figure 5.2). These rivers drain predominantly from the United States along the Gulf's northern margin, where the Mississippi and numerous smaller rivers enter the Gulf, and from Mexico's Yucatan Peninsula along the Gulf's southern margin, where Mexico's two largest rivers, the Grijalva and Usumacinta, discharge. Most dominant of these is the Mississippi River, whose plume water has been seen to extend as far southwest as the US-Mexican border, and as far southeast as Miami, FL on the Atlantic coast (Ortner et al. 1995).

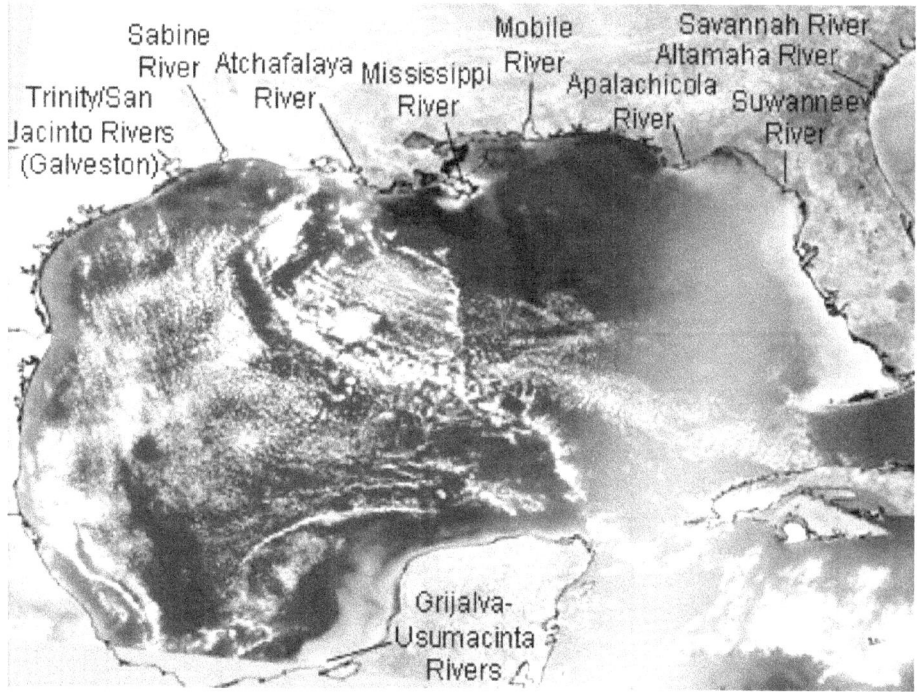

Figure 5.1. Some major river systems in the Gulf of Mexico overlaid on a 23 March 2001 MODIS (Terra) true color image. (Modified from Lohrenz and Verity, 2006).

Subregions of the GoM coast

Broadly, the GoM may be divided into four parts: the West Florida Shelf (WFS) located in the northeast part; the Northern GoM, which includes the Alabama and Mississippi (AL-MS) shelf and the Louisiana-Texas (LA-TX) shelf; the East Mexico shelf; and the Yucatan-Campeche shelf (Boicourt et al. 1998). These regions experience similar insolation and episodic storm forcing, but differ in their bathymetry, terrestrial freshwater input, and interaction with the currents of the open GoM.

The WFS is broad and shallow and consists of mostly carbonate-deposit sediments (Bianchi et al. 1999). As a result of its bathymetry, it is mostly isolated from the offshore eddies of the GoM, and circulation of the inner shelf is driven by local wind forcing, while the Loop Current (LC) dominates circulation at the outer shelf. Results from numerical modeling and observation have demonstrated distinct winter and summer seasonal patterns in shelf circulation (Lohrenz and Verity, 2006 and references therein). Winter circulation was characterized by offshore surface transport and coastal upwelling, while the summer circulation was characterized by northwestward-directed flow, onshore surface transport, and coastal downwelling (Yang and Weisberg, 1999, in Lohrenz and Verity, 2006). Bottom bathymetry and coastline morphology additionally influenced localized patterns in circulation. Currents in the outer shelf are, at times, subject to the influence of the LC southward flow. Freshwater inputs have localized impacts in the inner shelf the WFS, and, in concert with upwelling, may contribute to regions of enhanced productivity along the shelf break (He and Weisberg, 2002). While there are no major sources of freshwater input to the WFS, there have been observations of episodic transport of Mississippi discharge reaching the WFS (Del Castillo et al. 2001)

Figure 5.2. Gulf of Mexico and its estuaries (From Bianchi et al. 1999).

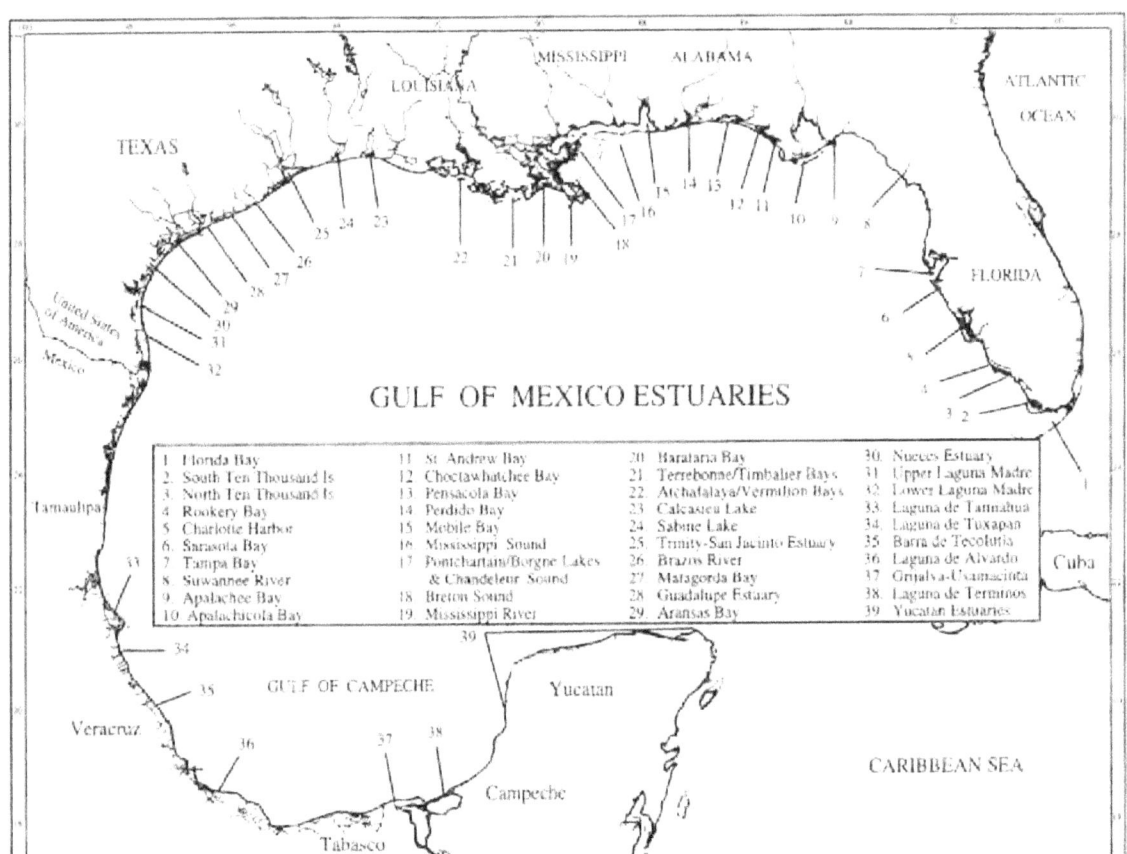

and even being transported out of the Gulf through the Straits of Florida (Ortner et al. 1995; Gilbert et al. 1996; Hu et al. 2005).

The Northern GoM is a broad, river-dominated margin whose seafloor consists mostly of terrigenous clastic sediments (Bianchi et al. 1999). Most of the freshwater comes from the Mississippi-Atchafalaya River system, although the Mobile River also provides a significant amount of freshwater to the eastern portion of this region. The plumes from the Mississippi Delta separate quickly from the bottom and spread buoyantly into deep water. The plume from the Mobile Bay behaves similarly but with significant mixing within the bay. The Atchafalaya River discharges into a broad shallow bay and mixes with seawater before moving offshore. The mechanisms by which dispersion of freshwater occurs across the coastal current fronts are not yet well understood. Currents in the northern GoM are forced by buoyancy, supplied mainly by the river discharge and supplemented by summer warming and episodic storms. Over the AL-MS shelf, the mean flow is westward. On the west side of the Mississippi Delta, the mean flow is eastward. Mean nearshore flow over the west LA-TX shelf is downcoast during all but summer months. The AL-MS shelf often is influenced by the LC while the LA-TX shelf is not strongly affected by the LC proper.

The Eastern Mexico shelf, from Texas to Campeche Bay, is narrow (50-100 km) and receives little direct freshwater input from the few small rivers that drain into the Gulf there (Figure 5.2). Despite this, the sediments of this region are mostly terrigeneous clastic deposits, most likely originating from the rivers from the northwest and southeast (Bianchi et al. 1999). In the Eastern Mexico shelf, currents flow southward during fall and winter, and reverse and flow to the northwest (towards Texas) during spring and summer.

The broader (over 200 km) Yucatan-Campeche shelf receives the discharge from Mexico's two largest rivers, the Grijalva and Usumacinta. In the Yucatan and Campeche shelf, the coastal flow is westward and relatively constant. As a result, little of the riverborne terrigenous sediments settle here, and the sediments are instead made up primarily of carbonate deposits (Bianchi et al. 1999).

Carbon Cycling on the Gulf Coast

The carbon cycle in the GoM is distinctly different between the river-dominated (Figure 5.3) and non-river-dominated regions. In the former, rivers deliver large amounts of POC, DOC, and sediments to ocean margins, and it is important to know the fractions of riverine POC and DOC recycled in the water column and surface sediment, exported to open ocean, and permanently buried in the sediments. As is the case for most river-dominated margins, we do not know these fractions well in the GoM's river-influenced shelves. For the Mississippi and Atchafalaya River deltas, sediment carbon content and stable carbon isotope ratios suggest that only 10% of these rivers' POC is preserved within the deltaic, shelf, and upper slope sediments, with the remainder either respired in sediments or water column or dispersed farther afield (Eadie et al. 1994).

Sediment dynamics in river-dominated margins have key implications for carbon diagenesis. Large lithogenic particles may serve as ballast materials for terrigeneous and marine C alike (Trefry et al. 1994), and flocculation and aggregation processes that occur in salinity gradients additionally accelerate sedimentation of organic materials (Dagg et al. 2004). High overall sedimentation rates in deltaic settings lead to longer-term organic carbon burial (Figure 5.4). Patterns in short-term sediment deposition may be distinctly different from long-term accumulation, an indication that river-borne sediments are highly mobile and

Figure 5.3. Approximate carbon budget for river-dominated regions of the Gulf of Mexico continental shelf. Uncertainties in respiration and shelf-ocean exchange preclude a rigorous budgeting.

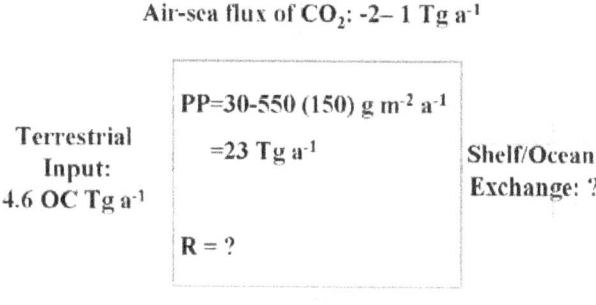

Air-sea flux of CO$_2$: -2– 1 Tg a^{-1}

Terrestrial Input: 4.6 OC Tg a^{-1}

PP=30-550 (150) g m^{-2} a^{-1} =23 Tg a^{-1}

R = ?

Shelf/Ocean Exchange: ?

Shelf Area (US): 1.56 x 10^5 km^2

Burial: .5-1.0 Tg a^{-1}

subject to episodic resuspension and redistribution (Wiseman et al. 1999).

Nutrient-rich river water also supports high levels of primary production in the Gulf of Mexico (Lohrenz et al. 1997). In the Mississippi River plume and nearby area, Redalje et al. (1994) measured particulate organic matter (POM) export flux at the base of the mixed layer together with surface primary production in both plume and shelf regions. Based on this work, annual POC export flux was 340 gC m^{-2} yr^{-1} out of an annual total PP of ~1230 gC m^{-2} yr^{-1}. Sediment trap δ^{13}C values also indicate that riverine POM did not make a large contribution to suspended materials in much of the plume and shelf waters (Redalje et al. 1994). This is consistent with the conclusion drawn from δ^{13}C values observed in surface sediments (i.e., 18 and 29% terrestrial contribution respectively at two sites; Eadie et al. 1994).

The high rate of production and deposition of organic matter (Lohrenz et al. 1997; Turner et al. 1998) produces a multivorous food web consisting of a mixture of micro- and picoplankton that supports higher consumers more efficiently (Pomeroy et al. 2000). River plumes also deliver terrigeneous DOC and POC, and waters underlying river plumes are perhaps sites of strong net respiration (e.g., Aller 1998; Aller and Aller, 2004), a key factor in the development of the hypoxic or 'dead' zones observed in the northern Gulf (Rabalais et al. 1999, 2001). McKee and Twilley (unpub., as cited in Dagg and Breed, 2003) estimated that 30% of the organic carbon deposited in the region immediately west of the delta is remineralized within four months. Farther to the west of the immediate deltaic-shelf environment, it was concluded that benthic recycling (0.2 gC m^{-2} d^{-1}) accounted for about 22% of the surface primary production (0.9 gC m^{-2} d^{-1}) (Dagg and Breed, 2003).

Many of the key features of the carbon cycle in these river-dominated regions are centrally dependent on an understanding of the rates of organic carbon degradation in the water column and sediments. The overall metabolic state of the Gulf Coast is not well constrained, however, with some studies suggesting large areas of net autotrophy (e.g., Lohrenz et al. 1997; Cai, 2003) and others implicating net heterotrophy (Walsh and Dieterle, 1994).

In contrast, the Eastern Mexico and West Florida shelves have minimal river influence and receive

nutrients mostly from subsurface water brought in by meanders and eddies of the LC, or by upwelling of deeper GoM water at the shelf break. The WFS is an oligotrophic system for most of the year. Chlorophyll-a concentrations in nearshore waters are typically 1 to 2 ug L^{-1} and rarely exceed 0.3 ug L^{-1} seaward of the 40-m isobath (Pomeroy et al. 2000). Episodic chlorophyll plumes during springtime have been noticed both in satellite imagery and during fieldwork (Darrow et al. 2003). Overall, the WFS is perhaps more like the South Atlantic Bight and has a microbial food web that is less efficient in supporting consumers (Pomeroy et al. 2000).

Net Air-Sea CO$_2$ Exchange

The net effect of these processes on air-sea CO$_2$ exchange is equivocal, as key identified forcings (riverine influences and storm forcings) can potentially drive both super- and undersaturation of surface-water pCO$_2$. Surface-water pCO$_2$ data are scarce in the GoM, with the exception of a few studies in the area of the Mississippi River Plume (Cai, 2003; Lohrenz and Cai, 2006), in the WFS (Wanninkhof et al. 1997; Millero et al. 2001; Clark et al. 2004), and a model prediction for the northern GoM (Walsh and Dieterle, 1994). We know of no studies of CO$_2$ distributions or fluxes in the East Mexico or Campeche-Yucatan shelves.

River inputs might potentially cause elevated pCO$_2$ as a result of accelerated respiration of river-derived

Figure 5.4. Long-term sediment accumulation rates based on 210Pb geochronologies. (From Wiseman et al. 1999).

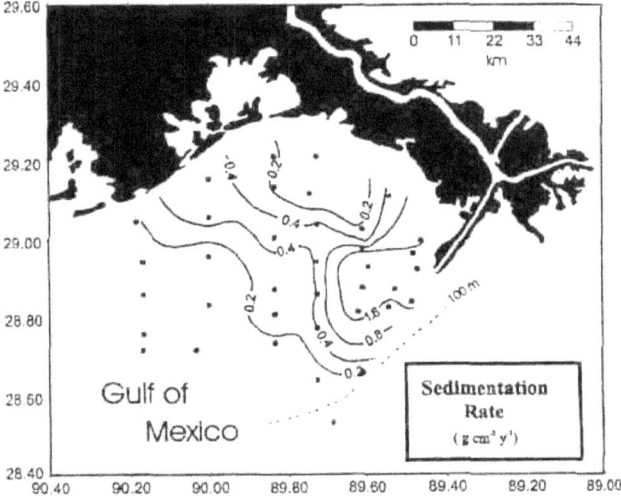

organic material (Aller, 1998; Aller and Aller, 2004). Alternatively, river-derived nutrients and freshwater-enhanced water-column stratification might fuel photosynthetic uptake of CO_2 and decreased surface pCO_2 levels. Preliminary field data suggest that surface pCO_2 value is highly variable in the Mississippi River plume-influenced area, although it is known that the lower river and the initial mixing area are a source of CO_2 to the atmosphere when turbidity is high, and the mid-field of the plume is a strong sink of atmospheric CO_2 (Cai, 2003; Lohrenz and Cai, 2006; see Figure 5.5). It is not known whether the entire LA-TX shelf (i.e., the freshwater-influenced area) is a sink or a source of CO_2, nor is the annual flux of the plume area known.

In non river-influenced settings such as the WFS, some studies have shown surface waters to be a source of CO_2 to the atmosphere (Millero et al. 2001; Clark et al. 2004). However, these sites were very nearshore

and thus potentially not representative of the broader WFS. Strong surface water pCO_2 under-saturation was measured in a more open shelf site after a spring bloom (Wanninkhof et al. 1997).

The other major forcing of the Gulf of Mexico, large episodic storms, can destratify the coastal water column, mobilizing metabolizable carbon from sediments and exposing high-pCO_2 waters to the sea surface, supporting release of CO_2 to the atmosphere. However, storms also cool surface waters and deliver nutrients that usually support post-storm plankton blooms, both of which support uptake of CO_2. On the other hand, generally strong heating in this shelf may help to drive pCO_2 towards supersaturation.

Based on the preliminary data and CO_2 flux direction in other similar systems (Cai et al. 2006), we speculate that the GoM is likely a source of CO_2 to the atmosphere. Resolving the magnitude of this flux

Figure 5.5(a). Dissolved inorganic carbon (DIC), total alkalinity (TAlk), and pH values measured in the Mississippi River plume in June 2003. Solid lines are regression lines and dashed lines are either the conservative mixing line (DIC) or lines (pH and pCO₂) predicted from conservation mixing. In the mixing calculation, river end-member DIC and TAlk (not shown) were used. Data from deep-water samples were included in the regression. This created an obvious offset in the pCO₂ case. Also, the river end-member value (900 µatm) was used to set the intercept in the pCO₂ regression. Similarly low pCO₂ was measured during underway surveys in the plume in summer 2004, 2005, and 2006 (Cai, unpublished).

Figure 5.5(b). Satellite-derived pCO₂ distribution in the shelf region near the Mississippi River delta derived from MODIS-Aqua imagery taken on 26 June 2003. The image shows a large area of low pCO₂ waters subject to the outflow of the Mississippi River that was consistent with in situ measurements (Figure 5.5a).

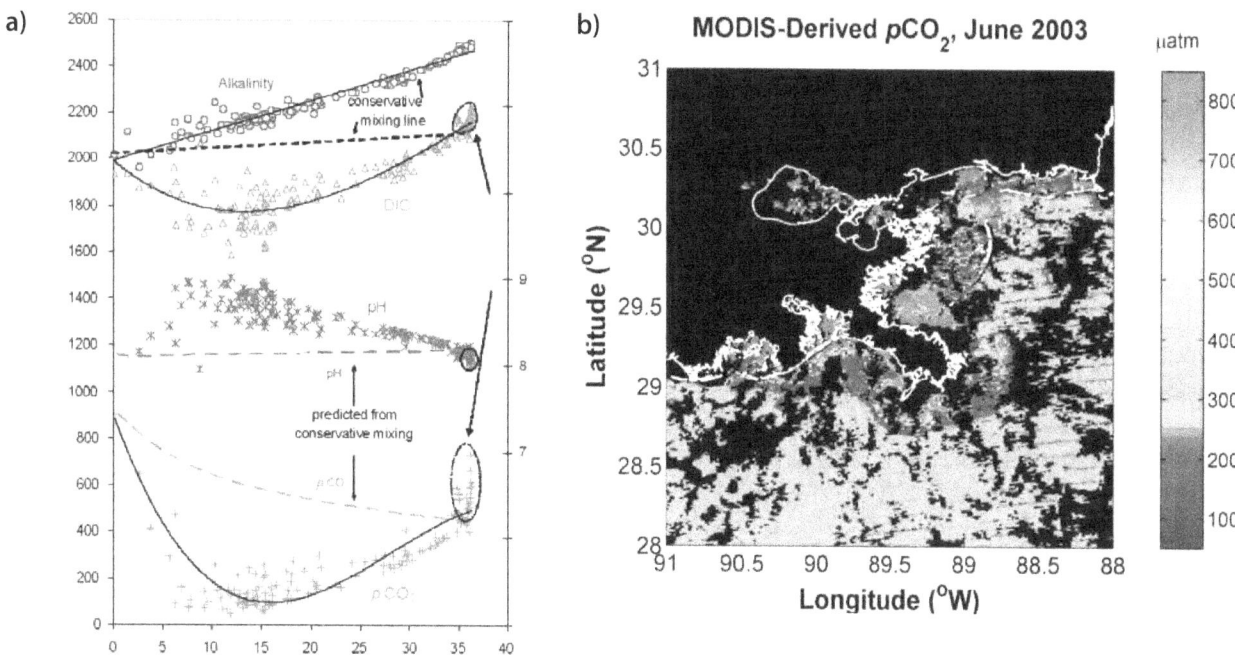

is critical, as evidenced from its disproportionate effect on the net CO_2 flux across the sea surface of North American coastal oceans (Chavez et al. 2007).

Historical Measurement Programs in the Gulf of Mexico

There has been no systematic, large-scale air-sea CO_2 flux measurement program in this area except for the few selected regional studies mentioned previously. NOAA will conduct a large-scale survey that will include measurements in the Gulf of Mexico and along the southeastern United States in summer 2007 (R. Wanninkhof, pers. comm.). The NOAA NECOP (Nutrient Enhanced Coastal Ocean Productivity) program has produced many useful and relevant results. They were summarized in a special issue of *ESTUARIES* in 1994 that has an introduction article and a series of papers. In addition, USGS has probably the world's most complete and detailed records on river water chemistry and sediment flux of the Mississippi River on their webpage. Finally, EPA has a comprehensive current program studying a large area between the Mississippi River and the Atchafalaya River (M. Murrell, pers. comm.). The EPA program explores the role of benthic and pelagic coupling on the development of the hypoxic zone. The CO_2 component is measured by W.-J. Cai's group.

Responses to Anthropogenic Forcing

It has been estimated that 25% of the net anthropogenic input of N to the entire drainage basin of the Mississippi River is eventually delivered via the Mississippi-Atchafalaya system to the Gulf of Mexico (Howarth et al. 1996). Long-term patterns in N flux show a substantial increase from 1967 through the early 1980's (Cai and Lohrenz, 2008), and a lack of consistent trend over the last two decades. The increase can be attributed, in part, to increased fertilizer use in the drainage basin (Turner and Rabalais, 1994), as well as to an increase in discharge (Cai and Lohrenz, 2008). Further details on river nutrient inputs can be found in Goolsby and Battaglin (2000, 2001), Goolsby et al. (1999), and Rabalais et al. (1999). It was found that inorganic carbon flux increased over the past half century in the Mississippi River and tributaries

(Raymond and Cole, 2003; Cai, 2003; Cai and Lohrenz, 2008). This increase was greater than that expected as a result of atmospheric CO_2 increase alone, and suggests that continental weathering rates have increased over the past half century in the Mississippi River Basin, perhaps as a result of changing land-use patterns.

These increases in nutrient fluxes are impacting the carbon cycle in river-dominated regions. For the region immediately west of the Mississippi River delta, Eadie et al. (1994) presented sediment isotopic information that suggests a productivity increase since 1950. In the immediate vicinity of river discharge, up to 70% of carbon burial is from terrestrial sources. Turner and Rabalais (1994) estimated an OC accumulation rate of 7.5 mol C m^{-2} yr^{-1} (or 90 gC m^{-2} yr^{-1}) during the 1980's. The increase in OC burial since the 1950s coincides with increased riverine N loading (Eadie et al. 1994).

Coastal wetlands, especially along the US central Gulf Coast, are being lost at an alarming rate due to factors such as sediment starvation, salt water intrusion, sea level rise, and canal dredging (e.g., Penland et al. 2005; Day et al. 2000; Turner, 1997). Recent devastating hurricanes along the Gulf Coast have reduced marsh surface area by ~10%, equivalent to an area >100 km^2 (e.g., Smith, 2005) and might be a tipping point in coastal erosion.

The Mississippi River discharge has increased in the last 50 years and has resulted in an increase in total alkalinity flux (Raymond and Cole, 2003; Cai, 2003). However, precipitation data and modeling for the past century (H. Tian, poster in NACP PI meeting, Colorado Springs, CO, 2007) suggest that in the first 50 years precipitation actually declined although it increased in the last 50 years of the last century. A climate model, based on a 3 to 4 °C temperature increase under doubled atmospheric CO_2 scenario, predicts that the Mississippi River discharge will decline in the following century due to increases in evapotranspiration that exceed increases in precipitation in the GoM and southeastern United States (Mulholland et al. 1997). Thus some apparently anthropogenic trends we have observed may not truly be anthropogenic or one-directional. Longer term observations and modeling are needed to predict future changes in river discharge and sediment delivery, and possible impacts on environments and carbon cycles.

References

Aller, J.Y., and R.C. Aller, 2004. Physical disturbance creates bacterial dominance of benthic biological communities in tropical deltaic environments of the Gulf of Papua. *Continental Shelf Research*, 24:2395.

Aller, R.C., 1998. Mobile deltaic and continental shelf muds as suboxic, fluidized bed reactors. *Marine Chemistry*, 61:143-155.

Berner, R.A., 1982. Burial of organic carbon and pyrite in the modern ocean: Its geological and environmental significance. *American Journal of Science*, 282:451-275.

Bianchi, T.S., J.R. Pennock, and R R. Twilley [eds.], 1999. *Biogeochemistry of Gulf of Mexico Estuaries*. John Wiley & Sons, Inc.

Boicourt, W.C., W.J. Wiseman, A. Valle-Levinson, and L.P. Atkinson, 1998. Continental shelf of the Southeastern United States and the Gulf of Mexico: In the shadow of the western boundary current. In: *The Sea, v.11* [A.R. Robinson and K.H. Brink (eds.)]. John Wiley & Sons, Inc.

Cai, W.-J., 2003. Riverine inorganic carbon flux and rate of biological uptake in the Mississippi River plume. *Geophysical Research Letters*, 30:1032.

Cai, W.-J., and S. Lohrenz, 2008. Carbon, nitrogen, and phosphorus fluxes from the Mississippi River and the transformation and fate of biological elements in the river plume and the adjacent margin. In: *Carbon and Nutrient Fluxes in Continental Margins: A Global Synthesis* [K.-K. Liu, L. Atkinson, R. Quiñones, and L. Talaue-McManus (eds)]. Springer-Verlag, New York, in press.

Cai, W.-J., M. Dai, and Y. Wang, 2006. Air-sea exchange of carbon dioxide in ocean margins: A province based synthesis. *Geophysical Research Letters*, 33:L12603, doi:10.1029/2006GL026219.

Chavez, F.P., T. Takahashi, W.-J. Cai, G. Friederich, B. Hales, R. Wanninkhof, and R. Feely, 2007. Coastal oceans. In: *The First State of the Carbon Cycle Report (SOCCR): The North American Carbon Budget and Implications for the Global Carbon Cycle*. [A.W. King, L. Dilling, G.P. Zimmerman, D.M. Fairman, R.A. Houghton, G. Marland, A.Z. Rose, and T.J. Wilbanks (eds.)]. A report by the U.S. Climate Change Science Program and the Subcommittee on Global Change Research, Washington, DC, pp. 157-166. Available at http://www.climatescience.gov/Library/sap/sap2-2/final-report/default.htm.

Clark, C.D., W.T. Hiscock, F.J. Millero, G. Hitchcock, L. Brand, W.L. Miller, L. Ziolkowski, R.F. Chen, and R.G. Zika, 2004. CDOM distribution and CO_2 production on the Southwest Florida Shelf. *Marine Chemistry*, 89:145-167.

Dagg, M.J., and G.A. Breed, 2003. Biological effects of Mississippi River nitrogen on the northern Gulf of Mexico—a review and synthesis. *Journal of Marine Systems*, 43:133.

Dagg, M., R. Benner, S. Lohrenz, and D. Lawrence, 2004. Transformation of dissolved and particulate materials on continental shelves influenced by large rivers: plume processes. *Continental Shelf Research*, 24:833-858.

Darrow, B.P., J.J. Walsh, G.A. Vargo, R.T. Masserini, K.A. Fanning, and J.Z. Zhang, 2003. A simulation study of the growth of benthic microalgae following the decline of a surface phytoplankton bloom. *Continental Shelf Research*, 23:1265-1283.

Day, J.W.J., G.P. Shaffer, L.D. Britsch, D.J. Reed, S.R. Hawes, and D.R. Cahoon, 2000. Pattern and process of land loss in the Mississippi Delta: a spatial and temporal analysis of wetland habitat change. *Estuaries*, 23:425-438.

Del Castillo, C.E., P.G. Coble, R.N. Conmy, F.E. Mueller-Karger, L. Vanderbloemen, and G.A. Vargo, 2001. Multispectral in situ measurements of organic matter and chlorophyll fluorescence in seawater: Documenting the intrusion of the Mississippi River plume in the West Florida Shelf. *Limnology and Oceanography*, 46:1836-1843.

Eadie, B.J., B.A. Mckee, M.B. Lansing, J.A. Robbins, S. Metz, and J.H. Trefrey, 1994. Records of nutrient enhanced coastal ocean productivity in sediments from the Louisiana continental shelf. *Estuaries*, 17:754-765.

Gilbert, P.S., T.N. Lee, and G.P. Podesta, 1996. Transport of anomalous low-salinity waters from the Mississippi River flood of 1993 to the Straits of Florida. *Continental Shelf Research*, 16:1065-1085.

Goolsby, D.A., and W.A. Battaglin, 2000. *Nitrogen in the Mississippi Basin—Estimating sources and predicting fluxes to the Gulf of Mexico*. USGS Fact Sheet 135-00. http://ks.water.usgs.gov/Kansas/pubs/fact-sheets/fs.135-00.html.

Goolsby, D.A., and W.A. Battaglin, 2001. Long-term changes in concentrations and flux of nitrogen in the Mississippi River Basin, USA. *Hydrological Processes*, 15:1209-1226.

Goolsby, D.A., et al. 1999. *Flux and Sources of Nutrients in the Mississippi-Atchafalaya River Basin: Topic 3 Report for the Integrated Assessment on Hypoxia in the Gulf of Mexico*. NOAA Coastal Ocean Program Decision Analysis Series No. 17. NOAA Coastal Ocean Program, Silver Spring, MD, 130 pp.

Gore, R.H., 1992. *The Gulf of Mexico*. Pineapple Press, Inc.

He, R., and R.H. Weisberg, 2002. West Florida Shelf circulation and heat budget for 1999 spring transition. *Continental Shelf Research*, 22:719-748.

Howarth, R.W., G. Billen, D. Swaney, A. Townsend, N. Jaworski, K. Lajtha, J. A. Downing, R. Elmgren, N. Caraco, T. Jordan, F. Berendse, J. Freney, V. Kudeyarov, P. Murdoch and Z. Zhao-Liang, 1996. Regional nitrogen budgets and riverine N & P fluxes for the drainages to the North Atlantic Ocean: Natural and human influences. *Biogeochemistry*, 35:75-139.

Hu, C., J.R. Nelson, E. Johns, Z. Chen, R.H. Weisberg, and F.E. Mueller-Karger, 2005. Mississippi River water in the Florida Straits and in the Gulf Stream off Georgia in summer 2004. *Geophysical Research Letters*, 32:L14606, doi:10.1029/2005GL022942.

Lohrenz, S.E., and W.-J. Cai, 2006. Satellite ocean color assessment of air-sea fluxes of CO_2 in a river-dominated coastal margin. *Geophysical Research Letters*, 33:L01601.

Lohrenz, S.E., and P.G. Verity, 2006. Regional Oceanography: Southeastern United States and Gulf of Mexico (2,W), Chapter 6 in *The Sea, Vol. 14A, The Global Coastal Ocean: Interdisciplinary Regional Studies And Syntheses* [A.R. Robinson and K.H. Brink (eds.)]. Harvard Press, Cambridge.

Lohrenz, S.E., G.L. Fahnenstiel, D.G. Redalje, G.A. Lang, X.G. Chen, and M.J. Dagg, 1997. Variations in primary production of northern Gulf of Mexico continental shelf waters linked to nutrient inputs from the Mississippi River. *Marine Ecology Progress Series*, 155:45-54.

Millero, F.J., W.T. Hiscock, F. Huang, M. Roche, and J.Z. Zhang, 2001. Seasonal variation of the carbonate system in Florida Bay. *Bulletin of Marine Science*, 68:101-123.

Mulholland, P.J., G.R. Best, C.C. Coutant, G.M. Hornberger, J.L. Meyer, P.J. Robinson, J.R. Stenberg, R.E. Turner, F. Veraherrera, and R.G. Wetzel, 1997. Effects of climate change on fresh water ecosystems of the South-Eastern United States and the Gulf Coast of Mexico. *Hydrological Processes*, 11:949-970.

Ortner, P.B., L.C. Hill, and S.R. Cummings, 1989. Zooplankton community structure and copepod species composition in the northern Gulf of Mexico. *Continental Shelf Research*, 9:387-402.

Ortner, P.B., T.N. Lee, P.J. Milne, R.G. Zika, M. Clarke, G.P. Podesta, P.K. Swart, P.A. Tester, L.P. Atkinson and W.R. Johnson, 1995. Mississippi River flood waters that reached the Gulf Stream. Journal of Geophysical Research, 100:13595-13601.

Penland, S., P.F. Connor, A. Beall, S. Feamley, and S.J. Williams, 2005. Changes in Louisiana shoreline: 1855-2002. Journal of Coastal Research, Sp. Iss., 44:7-39.

Pomeroy, L.R., J.E. Sheldon, W.M. Sheldon, J.O. Blanton, J. Amft, and F. Peters, 2000. Seasonal changes in microbial processes in estuarine and continental shelf waters of the south-eastern U.S.A. *Estuarine, Coastal and Shelf Science*, 51:415-428.

Rabalais, N.N., R.E. Turner, D. Justic, Q. Dortch, and W.J.J. Wiseman, 1999. *Characterization of Hypoxia: Topic 1 Report for the Integrated Assessment on Hypoxia in the Gulf of Mexico*. NOAA Coastal Ocean Program Decision Analysis Series No. 15. NOAA Coastal Ocean Program, Silver Spring, MD, 167 pp.

Rabalais, N.N., R.E. Turner, and W.J. Wiseman, 2001. Hypoxia in the Gulf of Mexico. *Journal of Environmental Quality*, 30:320-329.

Raymond, P.A., and J.J. Cole, 2003. Increase in the export of alkalinity from North America's largest river. *Science*, 301:88-91.

Redalje, D.G., S.E. Lohrenz, and G.L. Fahnenstiel, 1994. The relationship between primary production and the vertical export of particulate organic matter in a river impacted coastal ecosystem. *Estuaries*, 17:829-838.

Smith, G.J., 2005. Biological impacts of Hurricane Katrina on the Gulf Coast. *Geological Society of America National Meeting, 16-19 Oct 2005, Salt Lake City, Utah.* Paper LB1-4, http://gsa.confex.com/gsa/2005AM/finalprogram/session_16680.htm.

Trefry, J.H., S. Metz, T.A. Nelsen, R.P. Trocine, and B.J. Eadie, 1994. Transport of particulate organic carbon by the Mississippi River and the fate in the Gulf of Mexico. *Estuaries*, 17:839-849.

Turner, R.E., 1997. Wetland loss in the Northern Gulf of Mexico: multiple working hypotheses. *Estuaries*, 20:1-13.

Turner, R.E., and N.N. Rabalais, 1994. Coastal eutrophication near the Mississippi River delta. *Nature*, 368:619-621.

Turner, R.E., N. Qureshi, N.N. Rabalais, Q. Dortch, D. Justic, R.F. Shaw, and J. Cope, 1998. Fluctuating silicate:nitrate ratios and coastal plankton food webs. *Proceedings of the National Academy of Sciences*, 95:13048-13051.

Walsh, J.J., and D.A. Dieterle, 1994. CO_2 cycling in the coastal ocean. I - A numerical analysis of the southeastern Bering Sea with applications to the Chukchi Sea and the northern Gulf of Mexico. *Progress in Oceanography*, 34:335-392.

Wanninkhof, R., et al. 1997. Gas exchange, dispersion, and biological productivity on the west Florida shelf: Results from a Langrangian tracer study. *Geophysical Research Letters*, 24:1767-1770.

Wiseman, W.J. Jr., B. McKee, N.N. Rabalais, and S.P. Dinnel, 1999. Physical oceanography and sediment dynamics. In: *Nutrient Enhanced Coastal Ocean Productivity in the Northern Gulf of Mexico - Understanding the Effects of Nutrients on a Coastal Ecosystem* [W.J. Wiseman Jr., M.J. Dagg, N.N. Rabalais and T.E. Whitledge (eds.)]. NOAA Coastal Ocean Program Decision Analysis Series No. 14, Silver Spring, MD, pp 17-36.

Continental Margins of the Arctic Ocean and Bering Sea

Jackie Grebmeier
University of Tennessee- Knoxville

Nicholas R. Bates
Bermuda Institute of Ocean Sciences

Al Devol
University of Washington School of Oceanography

The Arctic Ocean, adjacent coastal seas such as the Chukchi, Beaufort, Barents, Kara, Laptev, and East Siberian Seas, and the subpolar Bering Sea contain 25% of the world's continental shelf areas (Figure 6.1). Strongly influenced by terrestrial freshwater inputs, the Arctic Ocean receives over 10% of total global river runoff, including that from four of North America's largest rivers. The margins of this system interact with the open ocean in three key ways. Low-salinity, high nutrient waters flow into the Bering Sea through the Aleutian Straits, where they strongly influence the Bering Sea shelf, but also flow through the Bering Strait and into the marginal Chukchi and Beaufort Seas. To the east, there is net transport of water through the system of passages and straits of the Canadian Arctic Archipelago into the Labrador Sea and North Atlantic Ocean. Along the Arctic coastlines of the Beaufort and Chukchi Seas and the Arctic Archipelago, exchange with the open Arctic Ocean can occur via upwelling of deep Arctic interior waters onto the shelves, and export of dense shelf waters—produced by brine rejection during sea-ice formation—to the deep Arctic Basin interior.

Unlike the other coastal oceans of the North American continent, the subpolar Bering and polar Arctic Ocean margins are highly influenced by extreme seasonal variations in insolation forcing that ultimately drive the advance and retreat of sea-ice and river discharge. Both freshwater inputs and sea-ice processes are important contributors to the hydrographic and biogeochemical properties of the polar mixed layer (PML) (e.g., Macdonald et al. 1989, 1995a,b, 2002; Anderson et al. 2004; Kadko and Swart, 2004; Cooper et al. 2005), strong water-column stratification (evident in the Canada and Eurasian Basins of the Arctic Ocean), and circulation patterns in the Arctic region. Sea-ice covers the Arctic margins and parts of the

Figure 6.1. Map of the Arctic Ocean with locations of the Chukchi Sea, Beaufort Sea, and Canada Basin. Minimum summertime (cyan) and maximum wintertime (purple) sea-ice extent taken from 1972-2002 SSMI data.

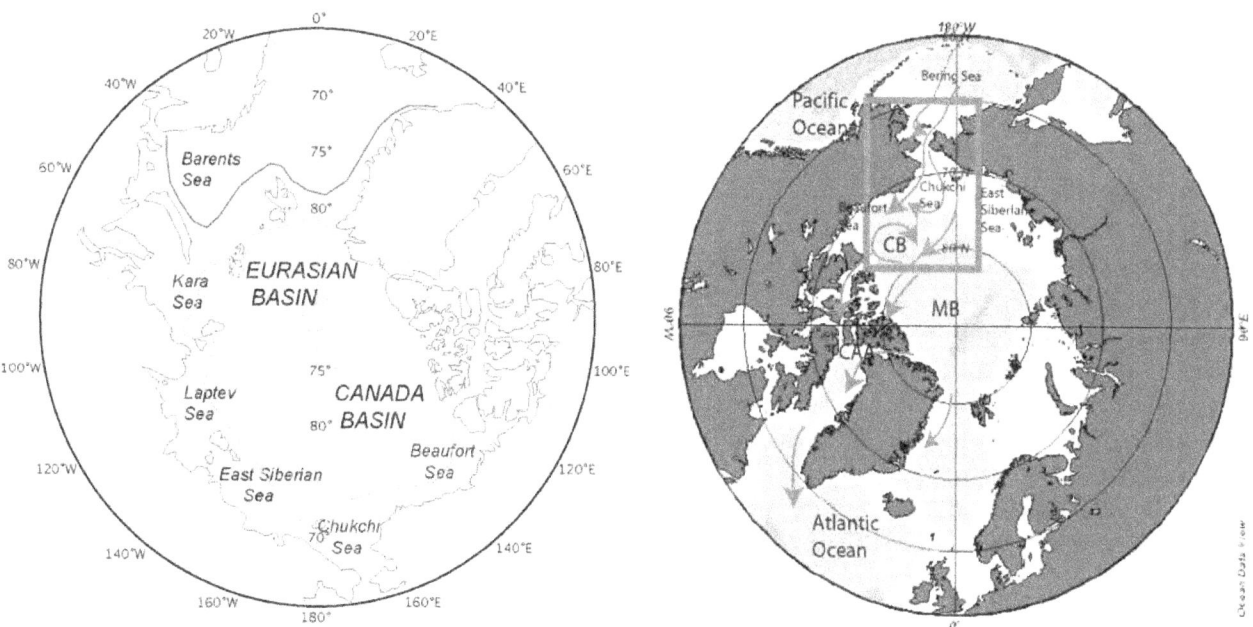

Bering Sea during wintertime months, but seasonally recedes from the coastlines in summertime.

This region has an important role in the global freshwater cycle (e.g., Aagaard and Carmack, 1989; Wijffels et al. 1992; Woodgate and Aagard, 2005) and Atlantic overturning circulation (e.g., Aagaard and Carmack, 1994; Walsh and Chapman, 1990; Mysak et al. 1990; Häkkinen, 1993; Wadley and Bigg, 2002). The flow of low-salinity, high-nutrient water from the North Pacific into the Arctic region through Bering Strait, combined with freshwater input and seasonal sea-ice melt and formation enhances the water-column stratification observed in the Canada and Eurasian Basins of the Arctic Ocean. Deepwater formation is a process unique to the Arctic among other North American margins, and represents a potential carbon sequestration pathway. The formation of dense seawater on the polar shelves due to wintertime cooling, sea-ice formation and brine rejection, and subsequent lateral advection of shelf waters into the central basin is an important pathway for transferring terrestrial and marine organic material from shelf seas to the deep ocean basin (Jones et al. 1990; Walsh and Chapman, 1990; Grebmeier, 1993; Grebmeier and Harvey, 2005; Benner et al. 2005).

The seasonal exposure of nutrient-rich surface water during summertime supports high rates of primary production and pelagic/benthic biomass, particularly in the Bering and Chukchi Seas. In contrast, the Arctic Ocean Basin is perennially oligotrophic with low rates of primary production (0.6–15 g C m^{-2} yr^{-1}; English, 1961; Cota et al. 1996; Wheeler et al. 1996; Moran et al. 1997; Gosselin et al. 1997; Chen et al. 2002; Anderson et al. 2003) and low pelagic/benthic biomass (Grebmeier et al. 2006a).

Subregions of the Arctic Margin

All areas of the Arctic margin experience extreme seasonal variations in insolation, sea-ice coverage, and inputs of terrestrial freshwaters and associated organic matter and nutrients. Based on differences in geomorphology, sea-ice processes, deepwater formation processes, and productivity regimes, however, this region can be divided into three distinct subregions: The Bering Sea, the Beaufort and Chukchi Seas, and the Arctic Archipelago including Hudson's and Ungava Bays.

The Bering Sea

The Bering Sea has an extensive (>1000 km length) and broad (>500 km width) shelf, with surface circulation consisting of relatively fresh, nutrient-rich waters derived from the North Pacific "funneling" towards Bering Strait (Figure 6.2). Sea ice is a factor in the Bering Sea, but to a lesser extent than in the other subregions. During the winter, ice covers much of the Bering shelf area, but the southward advance is constrained by the presence of relatively warm water in the central and southern Bering Sea. The Bering Sea, with less sea-ice coverage than the Arctic subregions, is subjected to the strong wintertime storm forcing of the North Pacific.

Chukchi and Beaufort Seas

The North American margins of the Chukchi and Beaufort Seas are exposed coastlines that are heavily impacted by inflow of cool, fresh, nutrient-rich Pacific water through the Bering Strait from the Bering Sea (Figure 6.3). This inflow is meandering and bifurcated in the broad and shallow (<100 m deep) continental shelf of the Chukchi Sea, but converges to flow eastward along the shelf break of the relatively narrow shelf of the Beaufort Sea toward the Canadian Arctic Archipelago (Coachman et al. 1975; Aagaard and Carmack, 1994; Overland and Roach, 1987; Björk, 1989). The mean inflow through the Bering Strait has an annual average of ~0.8 Sv, with higher flow in summer and lower flow in winter (Roach et al. 1995; Woodgate et al. 2005a,b). Water transiting Bering Strait is largely composed of warmer, fresher Alaskan Coastal Current (ACC) waters in the east (Paquette and Bourke, 1974), Bering Shelf water (central Bering Strait), and colder, saltier, more nutrient-rich water of the Anadyr Current in the west (Coachman et al. 1975). Bering Shelf and Anadyr water are thought to merge within the Chukchi Sea (Paquette and Bourke, 1974; Woodgate et al. 2005a,b; Codispoti et al. 2005). Another inflow of ~0.1 Sv (Woodgate et al. 2005a) to the Chukchi Sea is the intermittent Siberian Coastal Current from the East Siberian Sea through Long Strait (Weingartner et al. 1999). The four major outflows (~0.1-0.3 Sv each) from the Chukchi Sea into the Canada Basin of the central Arctic Ocean occur through Long Strait, Central Channel, Herald Valley, and Barrow Canyon (Paquette and Bourke, 1974; Woodgate et al. 2005a; Weingartner et al. 1998, 2005), although the outflow through Barrow Canyon can be as high as ~1 Sv (Münchow and Carmack, 1997).

Beaufort and Chukchi Sea coastal waters are highly influenced by seasonal sea-ice changes, and localized deep-water formation. Seasonal variations in sea-ice cover play a major role in shaping the Chukchi Beaufort Sea water masses and ecosystem. During the winter and much of the year, variable sea ice cover over the shelf areas results in an homogenized water column. During the summer, the sea ice retreats northward from the Bering Sea through Bering Strait, introducing open waters extending from the Chukchi and Beaufort Seas and beyond into the Canada Basin of the Arctic Ocean. On the Chukchi Sea shelf, local sea-ice melt transforms

water of Pacific Ocean origin to relatively warm, fresher Polar Mixed Layer (PML) water (upper 0 to 30 m, salinity typically <31; temperature >-1.5 °C). In the later part of the sea-ice free season, the temperature and salinity properties of the PML on the Chukchi Sea shelf widens from its wintertime temperature-salinity-density space (Woodgate et al. 2005a).

The Canadian Arctic Archipelago

The Canadian Arctic Archipelago (Figure 6.4) contains nearly 40,000 islands covering over one million square kilometers, with numerous shallow embayments and narrow straits and passages, and is geomorphologically distinct from the other Arctic/ Bering subregions. Two of North America's largest

Figure 6.2. The southeastern Bering Sea surface water circulation. Water flows across the shelf in the north (around St. Lawrence Island and through Bering Strait), but along the shelf in the southeast. Figure was modified from the BEST Science Plan (2004) and redrawn by N.R. Bates.

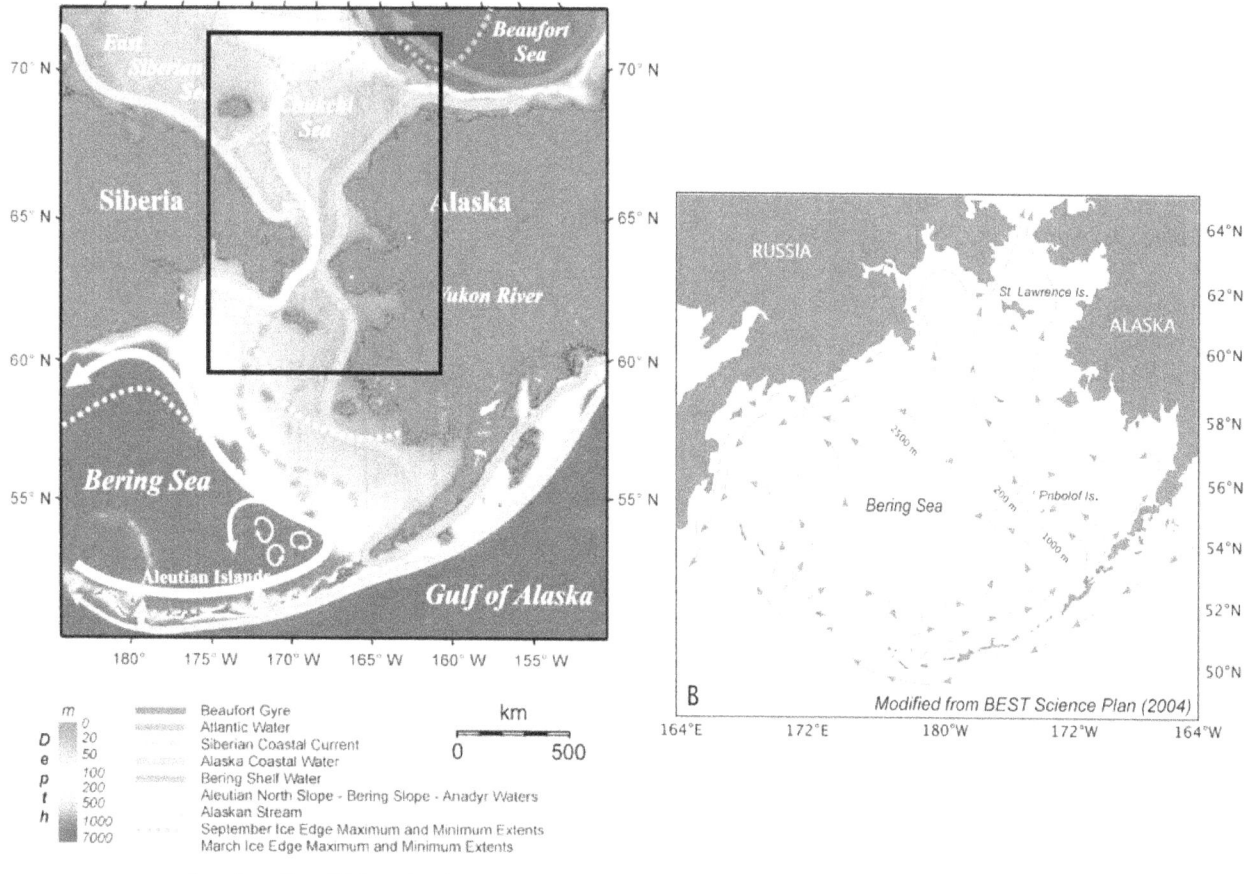

[courtesy Tom Weingartner]

rivers—the Nelson and Canniapiscau-Koksoak—drain into the Arctic Archipelago via the Hudson and Ungava Bays, which form the majority of the eastern shoreline with the mainland. The Alaskan Coastal Current and other inflows from the Canadian Arctic Basin deliver surface water through several passages along the north and west boundary of the archipelago, and water flows out through three main passages (Lancaster, Jones, and Smith Sounds) into the Labrador Sea. The oceanic character of these source waters shows less impact of the Pacific source chemistry, with lower nutrient concentrations than seen in the Bering or Chuckchi/Beaufort Sea subregions. Water transport through the sounds on the eastern border of the archipelago has a significant impact on freshwater exchange between Arctic and Atlantic Oceans, and this balance can affect North Atlantic deep water formation processes.

Carbon cycling

Carbon cycling in the Arctic margin is highly seasonal, a result of extreme changes in insolation, storm forcing, sea-ice coverage, and river discharge. Winter-time cross-shelf renewal of nutrients (Whitledge et al. 1986), followed by the springtime onset of stratification imposed by the influx of river water from the continent as well as by ice melt, combine with long hours of summer irradiance to make summertime coastal waters of the Bering, Beaufort, and Chukchi Seas highly productive marine ecosystems. The Bering Sea "green belt" (e.g., McRoy and Goering, 1974; Kinder and Coachman, 1978; Hansell et al. 1993; Springer and McRoy, 1993; Springer et al. 1996; Okkonen et al. 2004) extends northwards along the continental shelf, supporting a large population of seabirds, and marine mammals, >50% of commercial fish and shellfish landings in the United States, and traditional use of marine resources by the native peoples of Alaska and eastern Siberia. There are large gradients in primary production and pelagic (and benthic) biomass

Figure 6.3. Chukchi and Beaufort Sea mean circulation patterns.

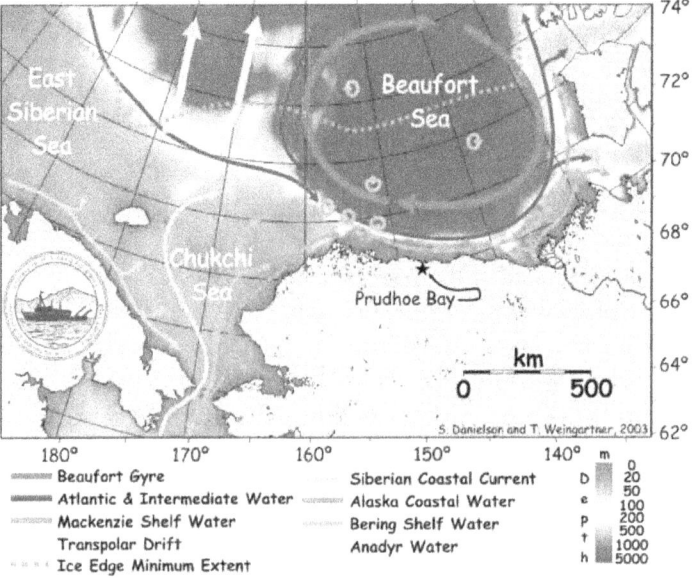

Figure 6.4. CAA map.

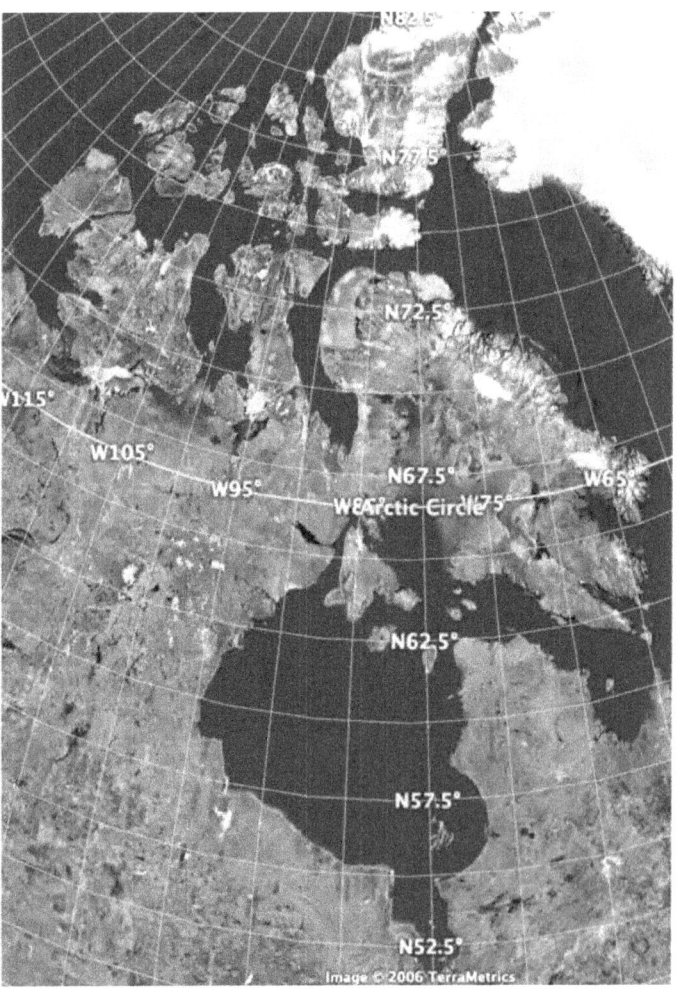

between the "green belt" of the shelf (Springer et al. 1996; Springer and McRoy, 1993; Grebmeier et al. 2006a) and oceanic basin (Kinder and Coachman, 1978).

During the brief "summertime" peak period of seasonal sea-ice melt and retraction toward the pole, the flow of nutrient-rich Pacific and Alaskan coastal waters through the Bering Strait into the Chukchi Sea from the Bering Sea supports a brief but intense photosynthetic season in the seasonally sea-ice free regions of the Chukchi and Beaufort Seas (Figure 6.5). Rates of water column ^{14}C primary and net community production (NCP) observed in the shallow coastal waters of the Chukchi Sea can be greater than 300 g C m^2 yr^{-1} (0.3-2.8 g C m^2 d^{-1}; Hameedi, 1978; Cota et al. 1996; Sambrotto et al. 1993; Hansell et al. 1993; Wheeler et al. 1996; Gosselin et al. 1997; Chen et al. 2002; Hill and Cota, 2005; Bates et al. 2005a; Walsh et al. 2005) (Figure 6.6). This primary production supports substantial pelagic and benthic biomass that, in turn, supports higher trophic levels (e.g., fish, marine mammals, seabirds) and important human socio-cultural and economic activities (Grebmeier and Harvey, 2005).

In the broad and shallow Bering and Chuckchi seas, much of the primary production is exported vertically to the seafloor (Moran et al. 2005) where there are high rates of benthic O_2 consumption (Figure 6). This probably results in a fairly tightly coupled recycling loop with recycled nutrients returned to the water column. However, denitrification in Arctic Ocean sediments may be a significant limitation on net primary production, especially in the productive Chukchi area where up to 50% (or more) of the combined nitrogen flux to the sediment may be denitrified. Also, it has been estimated that denitrification in the Arctic Ocean may represent of the order of 20% of the total global denitrification (Devol et al. 1997; Chang and Devol, in press) and thus play an important role in regulating global productivity. At the shelf break and along the narrower margin of the Beaufort Sea, significant suspended particulate organic matter (POM) is exported horizontally offshore into the interior of the Arctic Ocean Basin (Bates et al.

Figure 6.5. Satellite images of phytoplankton blooms following sea-ice retreat from the margins of the Chukchi and Beaufort Seas. Left panel courtesy Lou Codispoti, 2005; right panel from Hill and Cota, 2005.

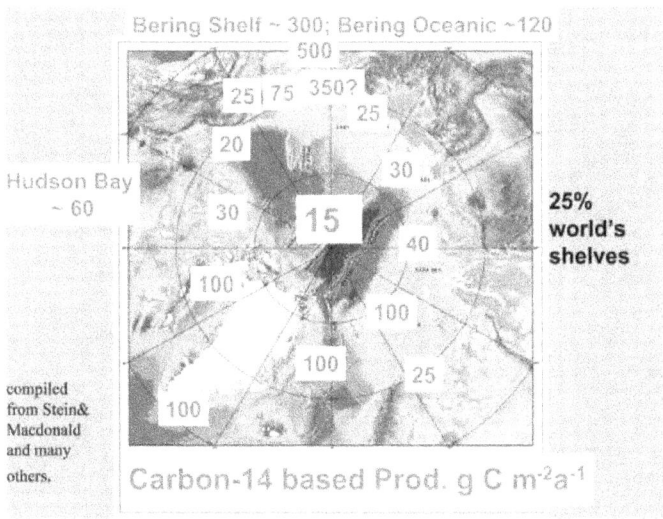

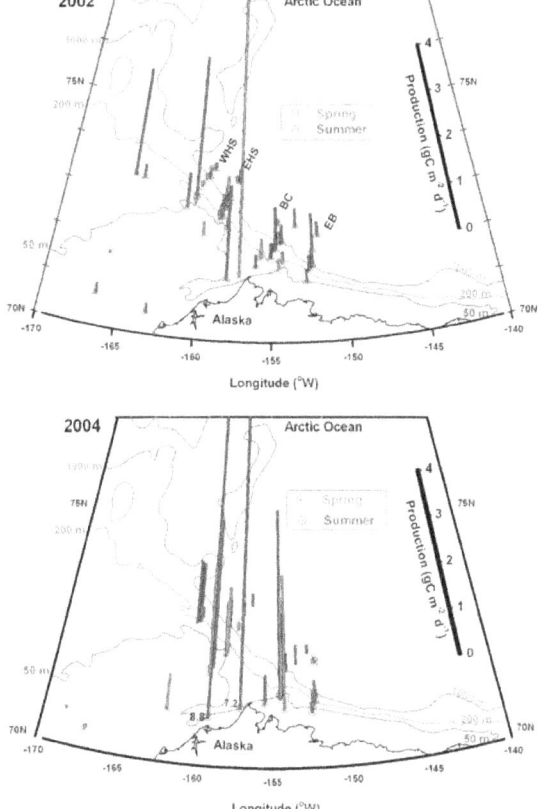

Figure 6.6. Estimates of primary productivity in Arctic margins based on a) historical compilation from Stein and MacDonald, and b) 2002 and 2004 results from the SBI project. Satellite images courtesy of G. Cota; phytoplankton photos courtesy of E. Sherr.

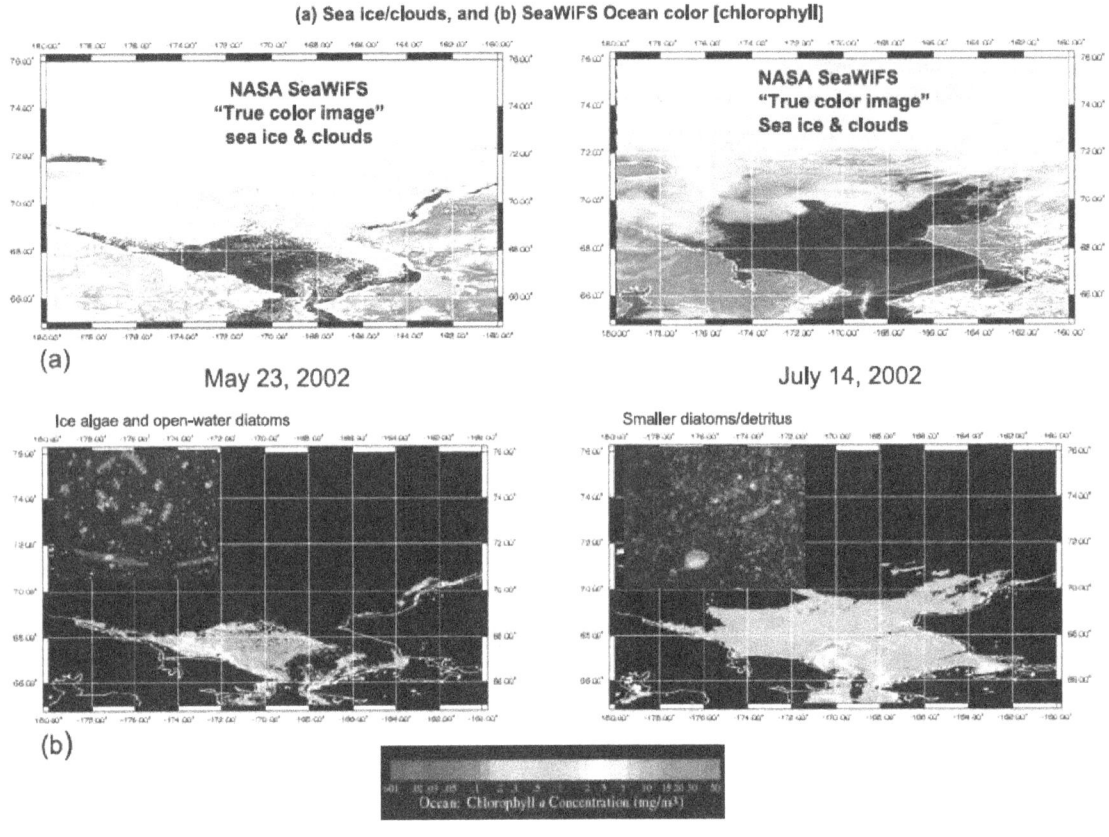

Figure 6.7. Distribution of a) benthic O₂ consumption rates, and b) benthic biomass in the Bering and Chukchi shelves. High export productivity and shallow water columns result in tight coupling between surface waters and benthos. From Grebmeier et al. (2006a).

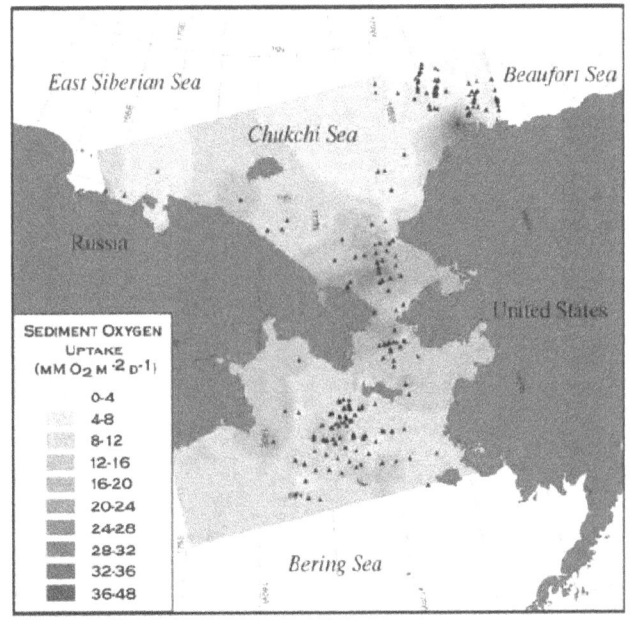
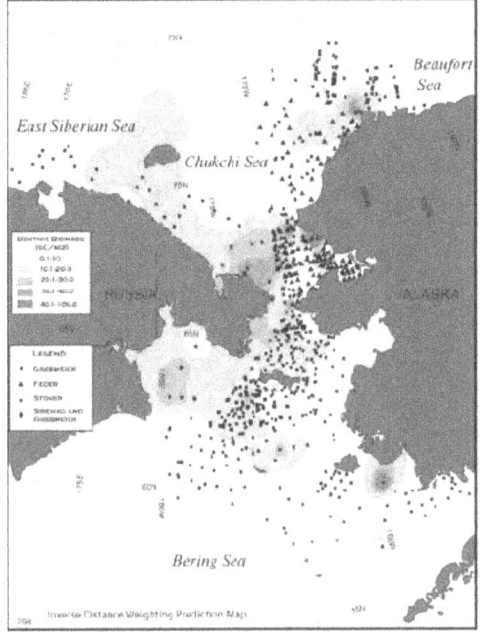

2005b) and subsequently delivered to greater depth. This, along with denitrification in shelf sediments, leads to depletion of nutrients in the ACC exiting the Beaufort Sea and flowing into the Arctic Archipelago. Both dissolved inorganic and organic carbon are also exported from the shelf into the Arctic Ocean Basin by mesoscale eddy transport (Mathis et al. 2007).

Of the three identified subregions in the Arctic/Bering margin, the carbon cycle of the Arctic Archipelago has probably received the least study. In the Arctic Archipelago, decreased nutrient content of inflowing waters and restricted exchange with deep waters places a greater emphasis on sea ice processes (Wassmann et al. 2004; Figure 6.8) and the effects of terrestrial freshwater inputs in shaping the carbon cycle. In particular, ice algae may play a greater role in the annual photosynthetic productivity in the Arctic Archipelago than elsewhere in the Arctic margin (Fortier et al. 2001). The direct loss of dense waters cascading over the sills of the Jones, Smith, and Lancaster Sounds into the Labrador Sea represents a unidirectional export pathway for dissolved carbon species.

In all subregions of this area, input of terrestrial dissolved and particulate organic carbon via river discharge plays a major role, with the Mackenzie and Yukon Rivers together contributing over 4 Tg C yr^{-1} (Telang et al. 1991). The signature of terrestrial particles is evident even far from river mouths in the chemical and isotopic composition of sedimentary organic carbon (Goñi et al. 2000; Belicka et al. 2002; Yunker et al. 2005; Figure 6.9). High concentrations of dissolved lignin phenols, unique biomarkers of land plants, are found throughout polar surface waters of the Arctic Ocean (Opsahl et al. 1999; Benner et al. 2004, 2005). The strong terrestrial signatures in organic matter of Arctic sediments and polar surface waters indicate this material is transported great distances and is resistant to decomposition. The extent to which terrestrial organic matter contributes to the carbon and energy requirements of biota in the water column and sediments is not well known. Nonetheless, this is a large source of carbon to these marginal settings and must be accounted for in comprehensive carbon-cycle studies.

Figure 6.8. Images of ice algae, an important factor in the carbon cycle in early spring.

Ice algae important in spring vs water column production in summer

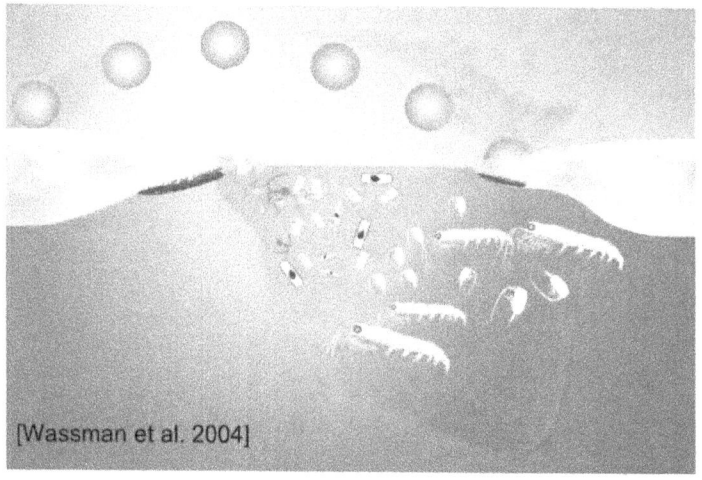

[Wassman et al. 2004]

[courtesy Rolf Gradinger]

Ice algae slower rates of decay compared to temperate diatoms –more "labile" material exported out of photic zone to fuel benthic production in shelf regions; the Pacific-influenced northern Bering, Chukchi, parts East Siberian and western Beaufort seas have both ice and open water production; are regions of high carbon production and export, thus benthic-rich systems

timing and location ice edge retreat critical for this relationship

Impact on Air-Sea CO_2 Exchange

The combination of cool waters, strong freshwater-driven stratification, abundant preformed nutrients, and rapid biological photosynthetic growth rates combine to make this region an apparently strong sink for atmospheric CO_2 in boreal spring and summer. Early studies suggested that the Bering Sea was a sink for atmospheric CO_2 (e.g., Kelley and Hood, 1971; Park et al. 1974; Codispoti et al. 1982, 1986; Walsh and Dieterle, 1994; Walsh et al. 1996). The high levels of primary production observed in the coastal Bering Sea "green belt" result in drawdown of inorganic nutrients and dissolved inorganic carbon (DIC) and pCO_2 (Codispoti et al. 1982, 1986). Estimates of CO_2 uptake in the Bering Sea range from 30 Tg C yr^{-1} (Cai et al. 2006) to 60 Tg C yr^{-1} (based on scaling Walsh and Dieterle's estimated area-specific air-sea flux by the area of the Bering Sea shelf).

There have been only a few studies of inorganic carbon cycling in the Chukchi Sea (Kelley, 1970;

Semiletov, 1999; Pipko et al. 2002; Murata and Takizawa, 2002; Yamamoto-Kawai et al. 2005; Kaltin and Anderson, 2005), and there are no published studies of the coastal Beaufort Sea or Canadian Arctic Archipelago. In the region of highest summertime primary or net community production in the Chukchi Sea, the photosynthetic fixation of CO_2 (Hill and Cota, 2005; Bates et al. 2005a), production of organic matter (Moran et al. 2005; Mathis et al. 2005, 2006), and the removal of inorganic and organic carbon from the surface waters led to very low seawater DIC (Bates et al. 2005a), and surface layer pCO_2 conditions (<100-150 µatm; Figure 6.10). This led to extremely high rates of air-to-sea CO_2 gas exchange (>90 mmol CO_2 m^{-2} d^{-1}). Seasonal (i.e., May to Sept.) and annual net air-to-sea CO_2 fluxes from the Chukchi Sea shelf were estimated at -27 ± 7 and 39 ± 7 Tg C yr^{-1}, respectively, representing the largest oceanic CO_2 sink in the marginal coastal seas adjacent to the Arctic Ocean (Bates, 2006). This compares to an annual net air-to-sea CO_2 flux of -22 Tg C yr^{-1} calculated indirectly for the

Figure 6.9. Distributions of ^{13}C content (‰) of organic carbon in surface sediments. There is enriched marine, diatom-based deposition in Anadyr and Bering Shelf water and a depleted terrestrial signal in nearshore Alaska coastal water and near the mouth of the Kolyma River. From Grebmeier et al. (2006a).

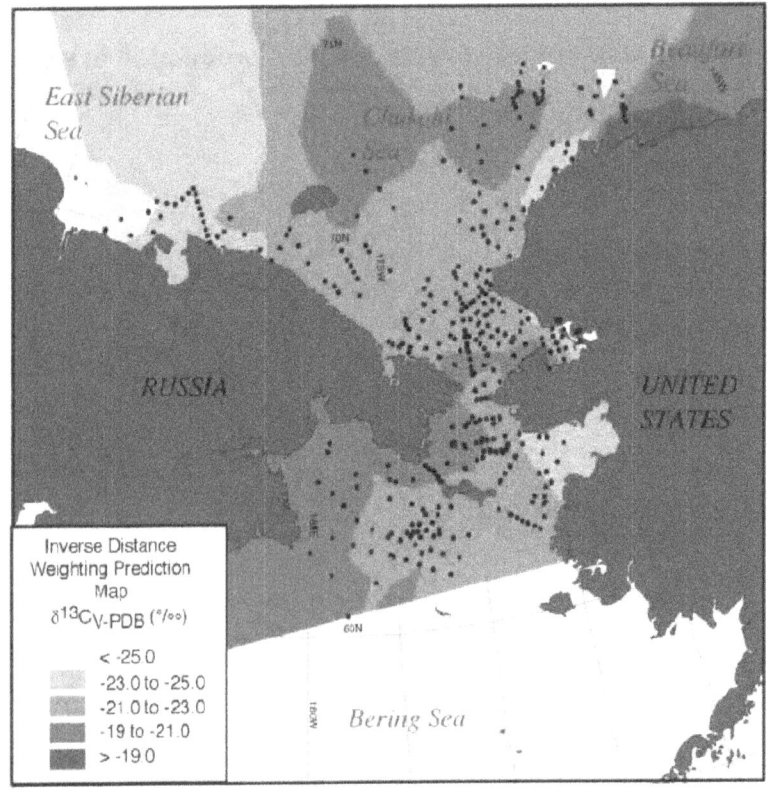

Figure 6.10. Distributions of seawater pCO₂ and air-sea CO₂ fluxes in the Chukchi Sea. a) Spring 2002 surface distributions of seawater pCO₂ (µatm) (HLY-02-01 cruise; 5 May-15 June 2002). Locations of CTD/water sampling stations in the Chukchi Sea, western Beaufort Sea, and Arctic Ocean are shown. Four sections were sampled from the Chukchi outer shelf into the Canada Basin of the Arctic Ocean, including: (1) West Hanna Shoal (WHS) transect; (2) East Hanna Shoal (EHS) transect; (3) Barrow Canyon (BC) transect, and; (4) East of Pt. Barrow (EB). Five stations were also sampled at Bering Strait. b) Summer 2002 surface distributions of seawater pCO₂ (µatm) (HLY-02-03 cruise; 17 July-26 August 2002). c) Spring 2004 surface distributions of seawater pCO₂ (µatm) (HLY-04-02 cruise; 17 May-21 June 2004). d) Spring 2002 air-to-sea CO₂ flux (mmol CO₂ m⁻² d⁻¹) (HLY-02-01 cruise; 5 May-15 June 2002). Contours (dashed line) indicate approximate distributions of sea-ice cover. e) Summer 2002 air-to-sea CO₂ flux (mmol CO₂ m⁻² d⁻¹) (HLY-02-03 cruise; 17 July-26 August 2002). f) Spring 2004 air-to-sea CO₂ flux (mmol CO₂ m⁻² d⁻¹) ((HLY-04-02; 17 May-21 Jun 2004). Figure from Bates (2006).

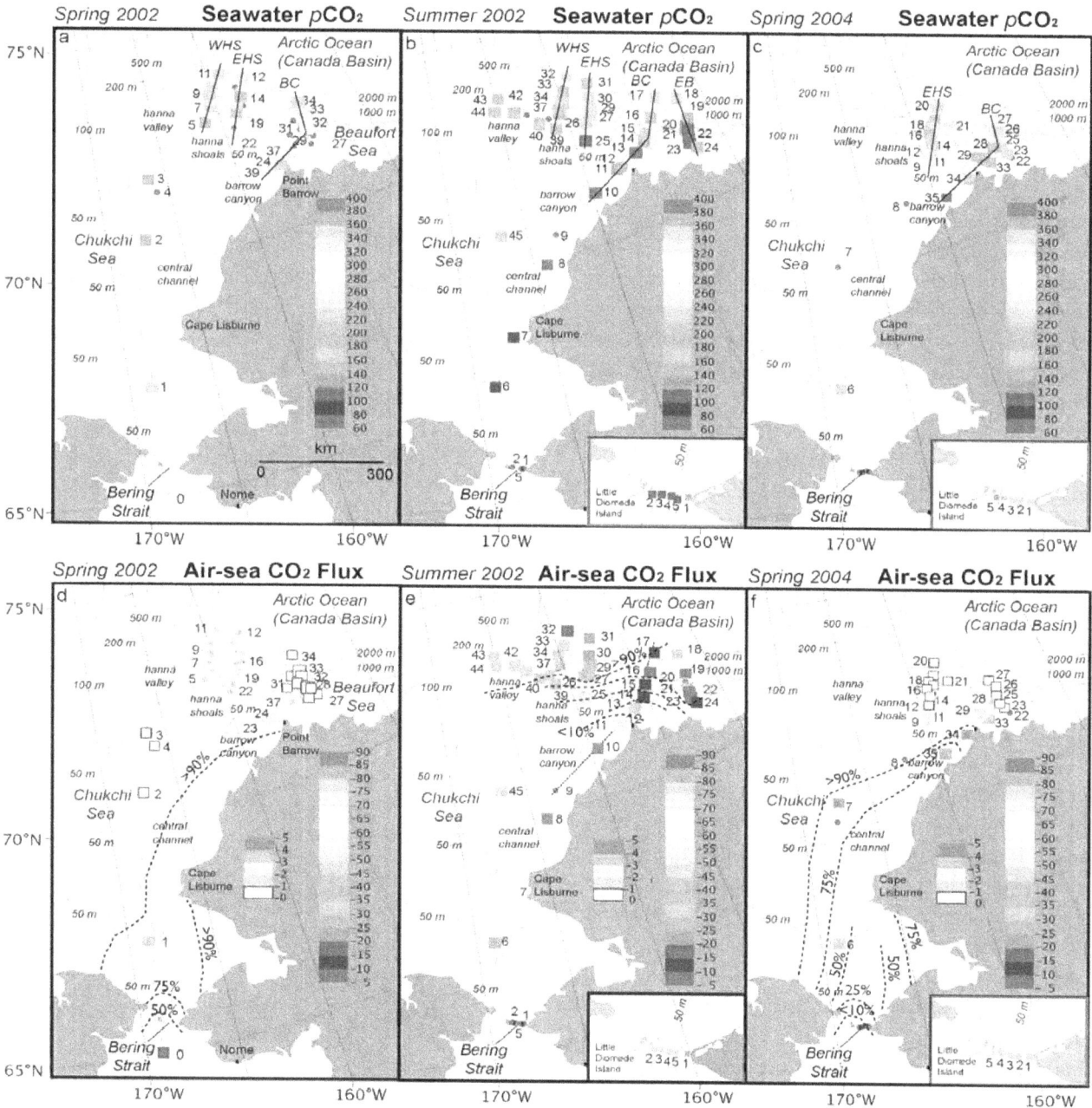

Chukchi Sea (Kaltin and Anderson, 2005; note that this estimate is also much higher than the Anderson et al. 1998 estimate) using a water and carbon transport mass balance. During sea-ice covered periods (sampled in early spring) mixed layer waters were weakly undersaturated with respect to the atmosphere, and the influx of CO_2 through air-sea gas exchange was low (<2 mmol CO_2 m^{-2} d^{-1}) due to smaller d pCO_2 values and suppression by sea-ice cover (~95-100%; Semiletov et al. 2004; Bates, 2006).

Anthropogenic Impacts

The Arctic and Bering Sea margins are, relatively, less impacted by direct anthropogenic activities than other North American margins. The region is largely unpopulated, and lightly impacted by agricultural and industrial activities. However, this is m ore than compensated for by the sensitivity of the region to long-term climate change. The region is particularly sensitive to natural long-term change and low-frequency modes of atmosphere-ocean-sea-ice forcing (e.g., Overland and Wang, 2005; Wang et al. 2005) such as the NAO and PDO. For example, the extent of sea-ice cover and ecosystem structure in the coastal Bering Sea undergoes significant interannual changes or regime shifts (e.g., Stabeno et al. 2002; Macklin et al. 2002; Hunt and Stabeno, 2002) that appear related to interannual changes in the low-frequency modes of the atmosphere such as the Arctic Oscillation (AO; Thompson and Wallace, 1998). The character of the marine ecosystem in the Bering Sea has undergone dramatic change in response to these regime shifts (e.g., Niebauer, 1998; Banse and English, 1999; Wylie-Echevarria and Ohtani, 1999; Stabeno et al. 2001, 2004; Hunt and Stabeno, 2002; Bond et al. 2003, Walsh et al. 2004; Grebmeier et al. 2006b), with cold-water, Arctic species being replaced by organisms more indicative of temperate zones. Previously infrequent coccolithophorid blooms have become regular features observed in the southeast Bering Sea (e.g., Stockwell et al. 2001; Merico et al. 2004, 2006; Figure 6.11), bringing associated elevated pCO_2 conditions (>400 µatm) and alkalinity depletions.

This demonstrates the sensitivity of the region to long-term climate shifts. Projected impacts of a warming climate are predicted to be more pronounced in the Arctic than elsewhere, and are described in depth in the recently-released State of the Arctic

Report (www.pmel.noaa.gov/pubs/PDF/rich2952/rich2952.pdf). Observed sea-ice extent and volume have decreased dramatically over the last few decades (e.g., Cavalieri et al. 2003; Rothrock and Zhang, 2005; Stroeve et al. 2005, 2007), and models predict more extreme ice loss in coming decades (Figure 6.12). This will drive complex atmosphere-ocean-sea-ice forcing interactions and feedbacks, influencing heat and freshwater budgets and exchanges. It will also have a complicated series of impacts on the ecosystem and the carbon cycle. Loss of sea ice will decrease the significance of ice algae, but may increase stratification and improve the physical conditions for mixed-layer phytoplankton blooms. It will increase the area of water exposed to the atmosphere, potentially increasing the wintertime Arctic atmospheric CO_2 sink (Bates et al. 2006), but will also expose the sea surface to wind forcing, potentially ventilating deeper, CO_2-rich waters to the surface in non-productive, low-insolation seasons. All of these projected changes suggest large perturbations to the carbon cycle of the margins of the Bering Sea and Arctic.

Historical Research Programs

Aside from the individual studies listed above and the Shelf-Basin Interactions program (see below),

Figure 6.11. Remotely sensed distribution of coccolithophorids over the south-eastern Bering Shelf. Figure was modified from the BEST Science Plan (2004).

there have been no large, coordinated carbon cycle research programs focused on the Arctic and Bering margins. Large historical and ongoing research programs include the Surface Heat Budget of the Arctic (SHEBA; http://www.eol.ucar.edu/projects/sheba/) project that was focused on physical processes in the Canadian Basin and had only ancillary carbon-relevant measurements (e.g., Yager, unpubl.; Sherr et al. 2003; Sherr and Sherr, 2003). The Arctic/Subarctic Ocean Fluxes West (ASOFW; http://asofw.apl.washington. edu/projects.html) program, under the 'A Study of Environmental Arctic Change' (SEARCH; http://psc. apl.washington.edu/search) project, includes several ongoing projects focusing on the physical transport of water through the passages of the Arctic Archipelago. The Study of the Northern Alaska Coastal System (SNACS; http://www.arcus.org/arcss/snacs) includes six ongoing research projects, two of which are focused specifically on terrestrial-marine carbon transports and transformations. The Shelf-Basin Interactions program (SBI; http://sbi.utk.edu/), focused on the Chukchi and Beaufort Seas, included extensive measurements of carbon standing stocks and rates of cycling in

the surface mixed layer, water column, and sea floor sediments (Figure 6.13). Carbon-relevant data was collected in each of these (e.g., oxygen, nutrient, and hydrographic property distributions; water transport rate estimates) but components of the SNACS and SBI projects are most relevant to carbon cycling in the Arctic margins. The SBI is an ongoing project that included field study years in 2002 through 2004, and is currently in a Synthesis and Modeling phase.

The NSF Bering Ecosystem Study (BEST; http:// www.arcus.org/Bering/index.html) is in its second year of a multi-year effort, and has expanded to collaborate with the North Pacific Research Board's Bering Sea Integrated Research Program (BSIERP; http://bsierp. nprb.org) to form a large, six-year multi-disciplinary ecosystem program for the Bering Sea. About 38 projects are supported through the BEST-BSIERP program, including several with direct relevance to carbon cycling on the Bering shelf. Finally, some carbon measurements are being made during IPY 2007/2008 as part of an 'Arctic Snapshot' Shelf Basin Exchange effort within the integrated Arctic Ocean Observing System (iAOOS) program under the direction of the

Figure 6.12. Sea ice changes in the past and predicted for the next several decades in the Arctic. From ACIA, 2004.

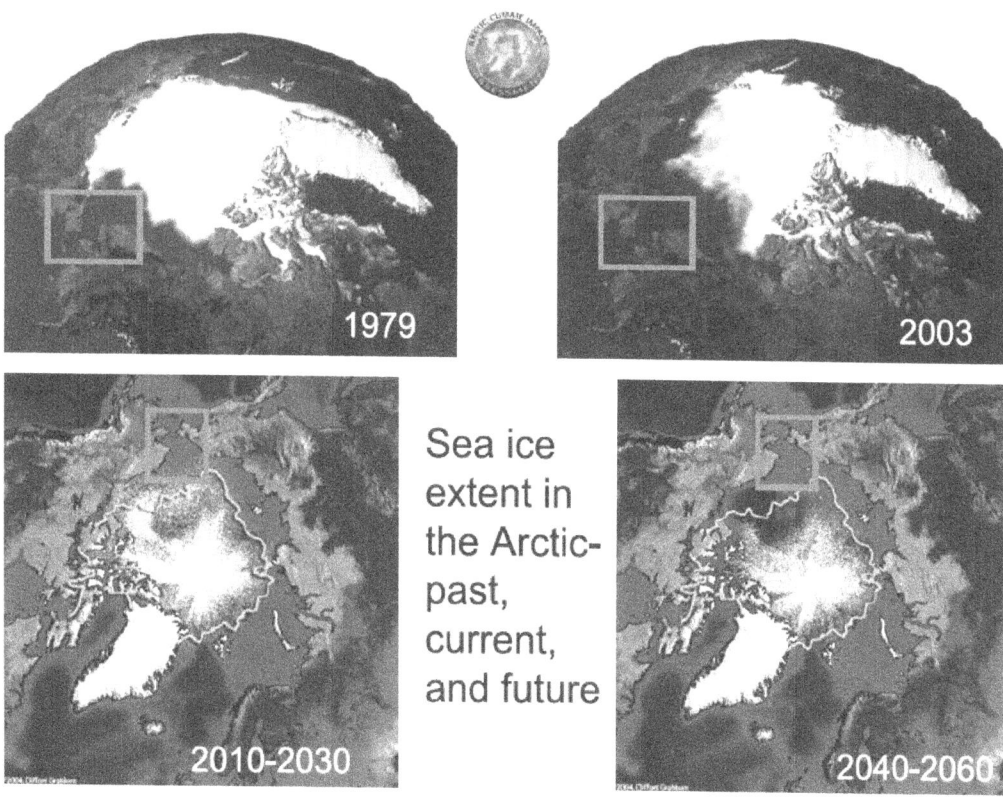

Arctic Ocean Sciences Board (http://www.aosb.org; Figure 6.14). The Science Plan for this effort (Working Group 5: Arctic Margins and Gateways) was developed as part of the Second International Conference on

Arctic Research Planning (ICARP II, 2007; http://www.icarp.dk) and many pertinent carbon cycling questions are discussed.

Figure 6.13. Studies that are a part of the SBI global change project are investigating the production, transformation, and fate of carbon at the shelf-slope interface in the northern Chukchi and Beaufort Seas, downstream of the productive shallow western Arctic shelves, as a prelude to understanding the impacts of a potential warming of the Arctic. Figure courtesy of Leif Anderson.

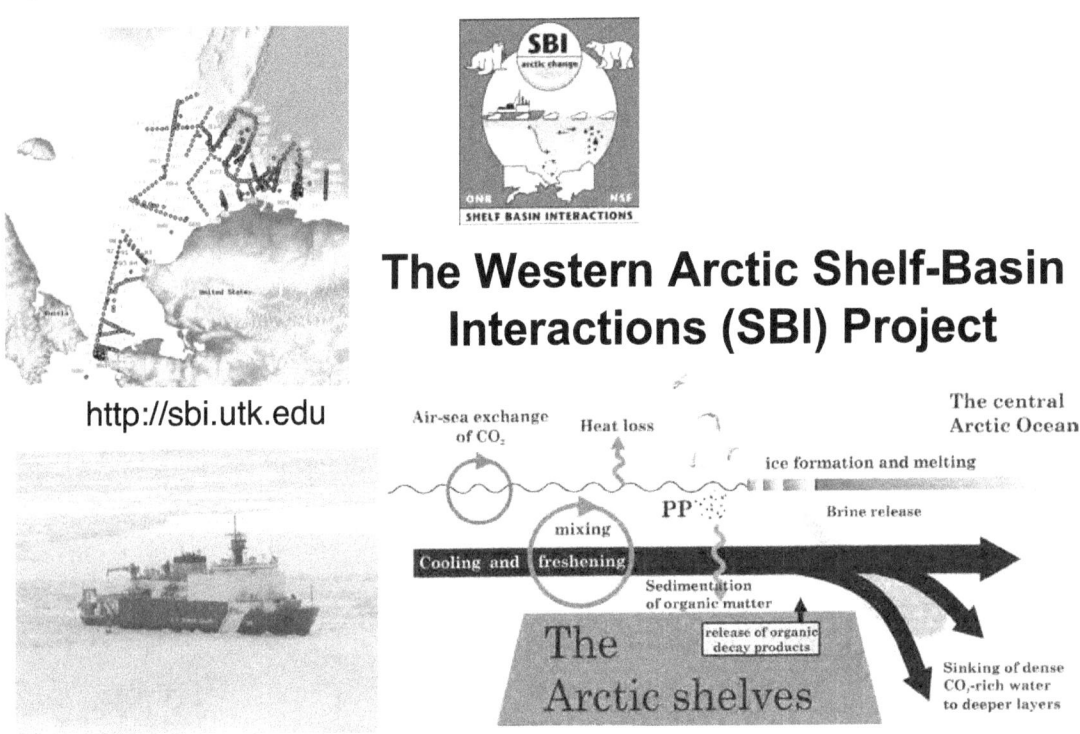

Figure 6.14. The AOSB/CLIC science bodies are currently undertaking an "Arctic Snapshot" of key shelf-basin exchange parameters at the shelf break and gateway complexes in the Arctic for the International Polar Year (IPY) in 2007-2009 through development of a synoptic network of collaborative international studies over a pan-Arctic scale as part of the integrated Arctic Ocean Observing System, iAOOS (http://www.aosb.org/pdf/iAOOS.pdf)html).

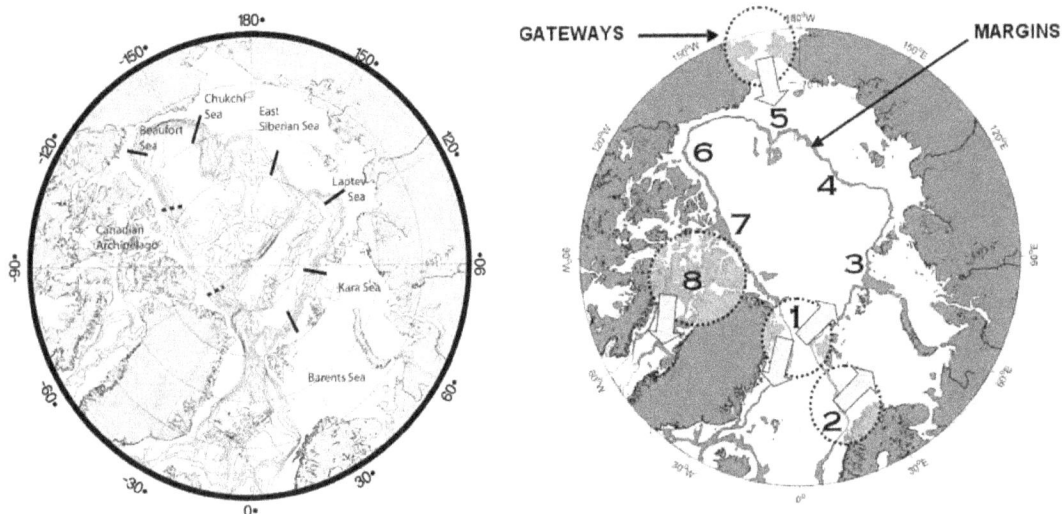

References

Aagaard, K., and E.C. Carmack, 1989. The role of sea ice and other fresh water in the Arctic circulation. *Journal of Geophysical Research*, 94:14485-14498.

Aagaard, K., and E.C. Carmack, 1994. The Arctic Ocean and climate: A perspective. In: *The Polar Oceans and Their Role in Shaping the Global Environment*. [O.M. Johannessen, R.D. Muench, and J.E. Overland (eds.)]. Geophysical Monograph 85. American Geophysical Union, pp. 5-20.

ACIA, 2004. *Impacts of a Warming Arctic: Arctic Climate Impact Assessment*. Cambridge University Press.

Anderson, L., K. Olsson, and M. Chierici, 1998. A carbon budget for the Arctic Ocean. *Global Biogeochemical Cycles*, 12:455-465.

Anderson, L.G., E.P. Jones, and J.H. Swift, 2003. Export production in the central Arctic Ocean evaluated from phosphate deficits. *Journal of Geophysical Research*, 108:3199, doi:10.1029/2001JC001057.

Anderson, L.G., S. Jutterstrom, S. Kaltin, E.P. Jones, and G.R. Bjork, 2004. Variability in river runoff distribution in the Eurasian Basin of the Arctic Ocean. *Journal of Geophysical Research*, 109:C01016.

Banse, K., and D.C. English, 1999. Comparing phytoplankton seasonality in the eastern and western subarctic Pacific and the western Bering Sea. *Progress in Oceanography*, 43:235-288.

Bates, N.R., 2006. Air-sea CO_2 fluxes and the continental shelf pump of carbon in the Chukchi Sea adjacent to the Arctic Ocean. *Journal of Geophysical Research*, 111:C10013, doi:10.129/2005JC003083.

Bates, N.R., M.H.P. Best, and D.A. Hansell, 2005a. Spatio-temporal distribution of dissolved inorganic carbon and net community production in the Chukchi and Beaufort Seas. *Deep-Sea Research II*, 52:3303-3323.

Bates, N.R., D.A. Hansell, S.B. Moran, and L.A. Codispoti, 2005b. Seasonal and spatial distributions of particulate organic matter (POM) in the Chukchi Sea. *Deep-Sea Research II*, 52:3324-3343.

Bates, N.R., S.B. Moran, D.A. Hansell, and J.T. Mathis, 2006. An increasing CO_2 sink in the Arctic Ocean due to sea-ice loss? *Geophysical Research Letters*, 33:L23609, doi:10.1029/2006GL027028.

Belicka, L.L., R.W. Macdonald, and H.R. Harvey, 2002. Sources and transport of organic carbon to shelf, slope, and basin surface sediments of the Arctic Ocean. *Deep-Sea Res. I*, 49:1463-1483.

Benner, R., B. Benitez-Nelson, K. Kaiser, and R.M.W. Amon, 2004. Export of young terrigenous dissolved organic carbon from rivers to the Arctic Ocean. *Geophysical Research Letters*, 31:L05305, doi:10.1029/2003GL019251.

Benner, R., P. Louchouarn, and R.M.W. Amon, 2005. Terrigenous dissolved organic matter in the Arctic Ocean and its transport to surface and deep waters of the North Atlantic. *Global Biogeochemical Cycles*. 19:GB2025, doi:10.1029/2004GB002398.

BEST, 2004. Bering Ecosystem Study (BEST) Science Plan. Arctic Research Consortium of the U.S., Fairbanks, AK, 82 pp.

Björk, G. 1989, A one-dimensional time-dependent model for the vertical stratification of the upper Arctic Ocean, *Journal of Physical Oceanography*, 19:52-67.

Bond, N.A., J.E. Overland, M. Spillane, and P. Stabeno, 2003. Recent shifts in the state of the North Pacific. *Geophysical Research Letters*, 30:2183.

Cai, W-J., M. Dai, and Y. Wang, 2006. Air-sea exchange of carbon dioxide in ocean margins: A province based synthesis. *Geophysical Research Letters*, 33:L12603, doi:10.1029/2006GL026219.

Cavalieri, D.J., C.L. Parkinson, and K.Y. Vinnikov, 2003. 30-Year satellite record reveals contrasting Arctic and Antarctic decadal sea ice variability. *Geophysical Research Letters*, 30:1970.

Chang, B.X., and A.H. Devol. Seasonal and spatial patterns of sedimentary denitrification rates in the shelf and slope sediments of the Chukchi Sea. *Deep-Sea Research II* (in press).

Chen, M., Y.P. Huang, L.D. Guo, P.H. Cai, W.F. Yang, G.S. Liu, and Y.S. Qiu, 2002. Biological productivity and carbon cycling in the Arctic Ocean. *Chinese Science Bulletin*, 47:1037-1040.

Coachman, L.K., K. Aagaard, and R.B. Tripp, 1975. *Bering Strait: The Regional Physical Oceanography*. University of Washington Press, Seattle.

Codispoti, L.A., G.E. Friederich, R.L. Iverson, and D.W. Hood, 1982. Temporal changes in the inorganic carbon system of the southeastern Bering Sea during spring 1980. *Nature*, 296:242-245.

Codispoti, L.A., G.E. Friederich, and D.W. Hood, 1986. Variability in the inorganic carbon system over the southeastern Bering Sea shelf during spring 1980 and spring-summer 1981. *Continental Shelf Research*, 5:133-160.

Codispoti, L., C. Flagg, and V. Kelly, 2005. Hydrographic conditions during the 2002 SBI process experiments. *Deep-Sea Research II*, 52:3199-3226.

Cooper, L.W., R. Benner, J.R. McClelland, B.J. Peterson, R.M. Holmes, P.A. Raymond, D.A. Hansell, J.M. Grebmeier, and L.A. Codispoti, 2005. The linkage between runoff, dissolved organic carbon, and the stable oxygen isotope composition of seawater and other water mass indicators in the Arctic Ocean. *Journal of Geophysical Research*, 110:G02013, doi:1029/2005JG000031.

Cota, G.F., L.R. Pomeroy, W.G. Harrison, E.P. Jones, F. Peters, W.M. Sheldon, and T.R. Weingartner, 1996. Nutrients, primary production and microbial heterotrophy in the southeastern Chukchi Sea: Arctic summer nutrient depletion and heterotrophy. *Marine Ecology Progress Series*, 135:247-258.

Devol, A.H., L.A. Codispoti, and J.P. Christensen, 1997. Summer and winter denitrification rates in the western Arctic shelf sediments. *Continental Shelf Research*, 17:1029-1050.

English, T.S., 1961. *Some Biological Oceanographic Observations in the Central North Polar Sea, Drift Station Alpha, 1957–1958.* Arctic Institute of North America Scientific Report 15.

Goñi, M.A., M.B. Yunker, R.W. Macdonald, and T.I. Eglinton, 2000. Distribution and sources of organic biomarkers in artic sediments from the Mackenzie River and Beaufort Shelf. *Marine Chemistry*, 71:23-51.

Gosselin, M., M. Levasseur, P.A. Wheeler, R.A. Horner, and B.C. Booth, 1997. New measurements of phytoplankton and ice algal production in the Arctic Ocean. *Deep-Sea Research II*, 44:1623-1644.

Grebmeier, J.M., 1993. Studies of pelagic-benthic coupling extended onto the Russian continental shelf in the Bering and Chukchi Seas. *Continental Shelf Research*, 13:653-668.

Grebmeier, J.M., and H.R. Harvey, 2005. The Western Arctic Shelf-Basin Interactions (SBI) project. An Overview. *Deep-Sea Research II*, 52:3109-3115.

Grebmeier, J.M., L.W. Cooper, H.M. Feder, and B.I Sirenko, 2006a. Ecosystem dynamics of the Pacific-influenced northern Bering and Chukchi Seas. *Progress in Oceanography*, 71:331-361.

Grebmeier, J.M., J.E. Overland, S.E. Moore, E.V. Farley, E.C. Carmack, L.W. Cooper, K.E. Frey, J.H. Helle, F.A. McLaughlin, and S.L. McNutt, 2006b. A major ecosystem shift in the northern Bering Sea. *Science*, 311:1461-1464.

Häkkinen, S., 1993. An Arctic source for the Great Salinity Anomaly: A simulation of the Arctic ice-ocean system for 1955-1975. *Journal of Geophysical Research*, 98:16397-16410.

Hansell, D.A., T.E. Whitledge, and J.J. Goering, 1993. Patterns of nitrate utilization and new production over the Bering-Chukchi shelf. *Continental Shelf Research*, 13:601-628.

Hameedi, M.J., 1978. Aspects of water column primary productivity in Chukchi Sea during summer. *Marine Biology*, 48:37-46.

Hill, V.J., and G.F. Cota, 2005. Spatial patterns of primary production in the Chukchi Sea in the spring and summer of 2002. *Deep-Sea Research II*, 52:3344-3354.

Hunt, G.L., and P. Stabeno, 2002. Climate change and the control of energy flow in the southeastern Bering Sea. *Progress in Oceanography*, 55:5-22.

ICARP, 2007. *Arctic Research: A Global Responsibility.* An Overview of the Second International Conference on Arctic Research Planning. 10-12 November 2005, Copenhagen, Denmark. McCallum Printing Group Inc., 44 pp. http://arcticportal.org/extras/portal/iasc/icarp/ICARP%20overview-0507-4.pdf.

Jones, E.P., D.M. Nelson, and P. Trequer, 1990. Chemical oceanography. In: *Polar Oceanography: Part B* [W.O. Smith (ed.)]. Academic Press, San Diego, pp. 407-476.

Kadko, D., and P. Swart, 2004. The source of the high heat and freshwater content of the upper ocean at the SHEBA site in the Beaufort Sea in 1997. *Journal of Geophysical Research*, 109:C01022, doi:10.1029/2002JC001734.

Kaltin, S., and L.G. Anderson, 2005. Uptake of atmospheric carbon dioxide in Arctic shelf seas: evaluation of the relative importance of processes that influence pCO_2 in water transported over the Bering-Chukchi Sea shelf. *Marine Chemistry*, 94:67-79.

Kelley, J.J., 1970. Carbon dioxide in the surface waters of the North Atlantic and the Barents and Kara Sea. *Limnology and Oceanography*, 15:80-87.

Kelley, J.J., and D.W. Hood, 1971. Carbon dioxide in the surface water of the ice-covered Bering Sea. *Nature*, 229:37-39.

Kinder, T.H., and L.K. Coachman, 1978. The front overlying the continental slope in the eastern Bering Sea. *Journal of Geophysical Research*, 83:4551-4559.

Macdonald, R.W., E.C. Carmack, F.A. McLaughlin, K. Iseki, D.M. Macdonald, and M.C. O'Brien, 1989. Composition and modification of water masses in the Mackenzie Shelf Estuary. *Journal of Geophysical Research*, 94:18057-18070.

Macdonald, R.W., D.W. Paton, E.C. Carmack, and A. Omstedt, 1995a. The fresh-water budget and under-ice spreading of Mackenzie River water in the Canadian Beaufort Sea based on salinity and O^{18}/O^{16} measurements in water and ice. *Journal of Geophysical Research*, 100:895-919.

Macdonald, R.W., E.C. Carmack, and D.W. Paton, 1995b. Using the delta O^{18} composition in landfast ice as a record of arctic estuarine processes. *Marine Chemistry*, 65:3-24.

Macdonald, R.W., F.A. McLaughlin, and E.C. Carmack, 2002. Fresh water and its sources during the SHEBA drift in the Canada Basin of the Arctic Ocean. *Deep-Sea Research I*, 49:1769-1785.

Macklin, S.A., G.L. Hunt, and J.E. Overland, 2002. Collaborative research on the pelagic ecosystem of the southeastern Bering Sea shelf. *Deep Sea Research II*, 49:5813-5819.

Mathis, J.T., D.A. Hansell, and N.R. Bates, 2005. Dissolved organic matter in the Chukchi Sea: Temporal and spatial variability. *Deep Sea Research II*, 52:3245-3258.

Mathis, J.T., R.S, Pickart, D.A. Hansell, D. Kadko, and N.R. Bates, 2007. Eddy transport of organic carbon and nutrients from the Chukchi Shelf into the deep Arctic basin. *Journal of Geophysical Research*, 112, C05011, doi:10.1029/2006JC003899.

McRoy, C.P., and J.J. Goering, 1974. The influence of ice on the primary productivity of the Bering Sea. In: *Oceanography of the Bering Sea with emphasis on renewable resources* [D.W. Hood and E.J. Kelley (eds.)]. University of Alaska, Fairbanks, Alaska, pp. 403-421.

Merico, A., T. Tyrrell, E.J. Lessard, T. Oguz, P.J. Stabeno, S.I. Zeeman, and T.E. Whitledge, 2004. Modelling phytoplankton succession on the Bering Sea role of climate influences and trophic interactions in generating *Emiliania huxleyi* blooms 1997-2000. *Deep Sea Research I*, 51:1803-1826.

Merico, A., T. Tyrrell, and T. Cokacar, 2006. Is there any relationship between phytoplankton seasonal dynamics and the carbonate system? *Journal of Marine Systems*, 59:120-142.

Moran, S.B., K.M. Ellis, and J.N. Smith, 1997. ^{234}Th/^{238}U disequilibrium in the central Arctic Ocean: Implications for particulate organic carbon export. *Deep Sea Research II*, 44:1593-1606.

Moran, S.B., R.P. Kelly, K. Hagstrom, J.N. Smith, J.M. Grebmeier, L.W. Cooper, G.F. Cota, J.J. Walsh, N.R. Bates, D.A. Hansell, and W. Maslowski, 2005. Seasonal changes in POC export flux in the Chukchi Sea and implications for water column-benthic coupling in Arctic shelves. *Deep Sea Research II*, 52:3427-3451.

Münchow, A., and E.C. Carmack, 1997. Synoptic flow and density observations near an arctic shelf break. *Journal of Physical Oceanography*, 27:1402-1419.

Murata, A., and T. Takizawa, 2002. Impact of a coccolithophorid bloom on the CO_2 system in surface waters of the eastern Bering Sea shelf. *Geophysical Research Letters*, 29:1547, doi:10.1029/2001GL013906.

Mysak, L.A., D.K. Manak, and R.F. Marsden, 1990. Sea ice anomalies in the Greenland and Labrador Seas during 1901-1984 and their relation to an interdecadal Arctic climate cycle. *Climate Dynamics*, 5:111-133.

Niebauer, H.J., 1998. Variability in Bering Sea ice cover as affected by a "regime shift" in the north Pacific in the period 1947-96. *Journal of Geophysical Research*, 103:27717-27737.

Okkonen, S.R., G.M. Schmidt, E.D. Cokelet, and P.J. Stabeno, 2004. Satellite and hydrographic observations of the Bering Sea 'Green Belt.' *Deep Sea Research II*, 51:1033-1051.

Opsahl, S., R. Benner, and R.M.W. Amon, 1999. Major flux of terrigenous dissolved organic matter through the Arctic Ocean. *Limnology and Oceanography*, 44:2017-2023.

Overland, J.E., and A.T. Roach, 1987. Northward flow in the Bering and Chukchi seas. *Journal of Geophysical Research*, 92:7097-7105.

Overland, J.E., and M. Wang, 2005. The Arctic climate paradox: The recent decrease of the Arctic Oscillation. *Geophysical Research Letters*, 32:L06701, doi:10.1029/2004GL021752.

Paquette, R.G., and R.H. Bourke, 1974. Observations on the coastal current of arctic Alaska. *Journal of Marine Research*, 32:195-207.

Park, P.K., L.I. Gordon, and S. Alvarez-Borego, 1974. The carbon dioxide system of the Bering Sea. In: *Oceanography of the Bering Sea* [D.W. Hood and J.J. Kelley (eds.)]. Institute of Marine Sciences, University of Alaska, Fairbanks, AK, pp. 107-147.

Pipko, I.I., I.P. Semiletov, P.Y. Tishchenko, S.P. Pugach, and J.P. Christensen, 2002. Carbonate chemistry dynamics in Bering Strait and the Chukchi Sea. *Progress in Oceanography*, 55:77-94.

Roach, A.T., K. Aagaard, C.H. Pease, S.A. Salo, T. Weingartner, V. Pavlov, and M. Kulakov, 1995. Direct measurements of transport and water properties through Bering Strait. *Journal of Geophysical Research*, 100:18443-18457.

Rothrock, D.A., and J. Zhang, 2005. Arctic Ocean sea ice volume: What explains its recent depletion? *Journal of Geophysical Research*, 110:C01002, doi:10.1029/2004JC002282.

Sambrotto, R.N., G. Savidge, C. Robinson, P. Boyd, T. Takahashi, D.M. Karl, C. Langdon, D. Chipman, J. Marra, and L. Codispoti, 2003. Elevated consumption of carbon relative to nitrogen in the surface ocean. *Nature*, 363:248-250.

Semiletov, I.P., 1999. Aquatic sources of CO_2 and CH_4 in the polar regions. *Journal of Atmospheric Sciences*, 56:286-306.

Semiletov, I.P., A. Makshtas, S.I. Akasofu, and E. Andreas, 2004. Atmospheric CO_2 balance: The role of Arctic sea ice. *Geophysical Research Letters*, 31:L05121, doi:10.1029/2003GL017996.

Sherr, B.F., and E.B. Sherr, 2003. Community respiration/production and bacterial activity in the upper water column of the central Arctic Ocean. *Deep Sea Research I*, 50:529-542.

Sherr, E.B., B.F. Sherr, P.A. Wheeler, and K. Thompson, 2003. Temporal and spatial variation in stocks of autotrophic and heterotrophic microbes in the upper water column of the central Arctic Ocean. *Deep-Sea Research I*, 50:557-571.

Springer, A.M., and C.P. McRoy, 1993. The paradox of pelagic food webs in the Northern Bering Sea. 3. Patterns of primary production. *Continental Shelf Research*, 13:575-599.

Springer, A.M., C.P. McRoy, and M.V. Flint, 1996. The Bering Sea Green Belt: shelf-edge processes and ecosystem production. *Fisheries Oceanography*, 5:205-223.

Stabeno, P.J., N.A. Bond, N.B. Kachel, S.A. Salo, and J.D. Schumacher, 2001. On the temporal variability of the physical environment over the southeastern Bering Sea. *Fisheries Oceanography*, 10:81-98.

Stabeno, P.J., N.B. Kachel, M. Sullivan, and T.E. Whitledge, 2002. Variability of physical and chemical characteristics along the 70-m isobath of the southeastern Bering Sea. *Deep Sea Research II*, 49:5931-5943.

Stockwell, D.A., T.E. Whitledge, S.I. Zeeman, K.O. Coyle, J.M. Napp, R.D. Brodeur, A.I. Pinchuk, and G.L. Hunt, 2001. Anomalous conditions in the south-eastern Bering Sea, 1997: Nutrients, phytoplankton and zooplankton. *Fisheries Oceanography*, 10:99-116.

Stroeve, J.C., M.C. Serreze, F. Fetterer, T. Arbetter, W. Meier, J. Maslanik, and K. Knowles, 2005. Tracking the Arctic's shrinking ice cover: Another extreme September minimum in 2004. *Geophysical Research Letters*, 32:L04501, doi:10.1029/2004GL021810.

Stroeve, J., M.M. Holland, W. Meier, T. Scambos, and M. Serreze, 2007. Arctic sea ice decline: Faster than forecast. *Geophysical Research Letters*, 34, L09501, doi:10.1029/2007GL029703.

Telang, S.A., R. Pocklington, A.S. Naidu, E.A. Romankevich, I.I. Gitelson, and M.I. Gladyshev, 1991. Carbon and mineral transport in major North American, Russian Arctic, and Siberian rivers: The Lawrence, the Mackenzie, the Yukon, the Arctic Alaskan Rivers, the Arctic basin rivers in the Soviet Union, and the Yenisei. In: *Biogeochemistry of Major World Rivers* [E.T. Degens, S. Kempe, and J.E. Richey (eds.)]. Wiley, Chichester, pp. 75-104.

Thompson, D.W.J., and J.M. Wallace, 1998. The Arctic Oscillation signature in the wintertime geopotential height and temperature fields. *Geophysical Research Letters*, 25:1297-1300.

Wadley, M.R., and G.R. Bigg, 2002. Impact of flow through the Canadian Archipelago and Bering Strait on the North Atlantic and Arctic circulation: An ocean modelling study. *Quarterly Journal of the Royal Meteorological Society*, 128:2187-2203.

Walsh, J.E., and W.L. Chapman, 1990. Arctic contribution to upper ocean variability in the North Atlantic. *Journal of Climate*, 3:1462-1473.

Walsh, J.J., and D.A. Dieterle, 1994. CO_2 cycling in the coastal ocean. I- A numerical analysis of the southeastern Bering Sea with applications to the Chukchi Sea and the northern Gulf of Mexico. *Progress in Oceanography*, 34:335-392.

Walsh, J.J., D.A. Dieterle, F.E. Muller-Karger, K. Aagaard, A.T. Roach, T.E. Whitledge, and D. Stockwell, 1996. CO_2 cycling in the coastal ocean. II. Seasonal organic loading of the Arctic Ocean from sources waters in the Bering Sea. *Continental Shelf Research*, 17:1-36.

Walsh, J.J., D.A. Dieterle, W. Maslowski, J.M. Grebmeier, and T.E. Whitledge, 2004. Decadal shifts in biophysical forcing of Arctic marine food webs: Numerical consequences. *Journal of Geophysical Research*, 109:C05031, doi:10.1029/2003JC001945.

Walsh, J.J., D.A. Dieterle, W. Maslowski, J.M. Grebmeier, T.E. Whitledge, M. Flint, I.N. Sukhanova, N. Bates, G.F. Cota, D. Stockwell, S.B. Moran, D.A. Hansell, and C.P. McRoy, 2005. A numerical model of seasonal primary production within the Chukchi/Beaufort Seas. *Deep Sea Research II*, 52:3541-3576.

Wang, J., M. Ikeda, S. Zhang, and R. Gerdes, 2005. Linking the northern hemisphere sea-ice reduction trend and the quasi-decadal arctic sea-ice oscillation. *Climate Dynamics*, 24:115-130.

Wassmann, P., E. Bauerfeind, M. Fortier, M. Fukuchi, B. Hargrave, B. Moran, T. Noji, E.-M. Nöthig, K. Olli, R. Peinert, H. Sasaki, and V. Shevchenko, 2004. Particulate organic carbon flux to the Arctic Ocean sea floor. In: *The Organic Carbon Cycle in the Arctic Ocean* [R. Stein and R.W. Macdonald (eds.)]. Springer, Berlin, pp. 101-138.

Weingartner, T.J., D.J. Cavalieri, K. Aagaard, and Y. Sasaki, 1998. Circulation, dense water formation and outflow on the northeast Chukchi Sea shelf. *Journal of Geophysical Research*, 103:7647-7662.

Weingartner, T.J., S. Danielson, Y. Sasaki, V. Pavlov, and M. Kulakov, 1999, The Siberian Coastal Current: A wind and buoyancy-forced arctic coastal current. *Journal of Geophysical Research*, 104:29697-29713.

Weingartner, T., K. Aagaard, R. Woodgate, S. Danielson, Y. Sasaki, and D. Cavalieri, 2005. Circulation on the North Central Chukchi Sea Shelf. *Deep-Sea Research II*, 52:3150-3174.

Wheeler, P.A., M. Gosselin, E. Sherr, D. Thibault, D.L. Kirchman, R. Benner, and T.E. Whitledge, 1996. Active cycling of organic carbon in the central Arctic Ocean. *Nature*, 380:697-699.

Whitledge, T.E., W.S. Reeburgh, and J.J. Walsh, 1986. Seasonal inorganic nitrogen distributions and dynamics in the southeastern Bering Sea. *Continental Shelf Research*, 5:109-132.

Wijffels, S.E., R.W. Schmitt, H.L. Bryden, and A. Stigebrandt, 1992. Transport of freshwater by the oceans. *Journal of Geophysical Research*, 22:155-162.

Woodgate, R.A., and K. Aagaard, 2005. Revising the Bering Strait freshwater flux into the Arctic Ocean. *Geophysical Research Letters*, 32:L02602, doi:10.1029/2004GL021747.

Woodgate, R.A., K. Aagaard, and T.J. Weingartner, 2005a. A year in the physical oceanography of the Chukchi Sea: Moored measurements from autumn 1990-1991. *Deep Sea Research II*, 52:3116-3149.

Woodgate, R.A., K. Aagaard, and T.J. Weingartner, 2005b. Monthly temperature, salinity and transport variability of the Bering Strait throughflow. *Geophysical Research Letters*, 32:L04601, doi:10.1029/2004GL021880.

Wyllie-Echeverria, T., and K. Ohtani, 1999. Seasonal sea ice variability and the Bering Sea ecosystem. In: *Dynamics of the Bering Sea* [T. Loughlin and K. Ohtani (eds.)]. University of Alaska Sea Grant Press, Fairbanks, Alaska, pp. 435-451.

Yamamoto-Kawai, M., N. Tanaka, and S. Pivovarov, 2005. Freshwater and brine behaviors in the Arctic Ocean deduced from historical data of delta O-18 and alkalinity (1929-2002 AD). *Journal of Geophysical Research*, 110:C10003, doi:10.1029/2004JC002793.

Yunker, M.B., L.L. Belicka, H.R. Harvey, and R.W. Macdonald, 2005. Tracing the inputs and fate of marine and terrigenous organic mater in Arctic Ocean sediments: A multivariate analysis of lipid biomarkers. *Deep Sea Research II*, 52:3478-3508.

The Laurentian Great Lakes

James T. Waples
Great Lakes WATER Institute
University of Wisconsin-Milwaukee

Brian Eadie
NOAA/GLERL

J. Val Klump
Great Lakes WATER Institute
University of Wisconsin-Milwaukee

Margaret Squires
Biology Department
University of Waterloo, Ontario, Canada

James Cotner
Department of Ecology, Evolution, and Behavior
University of Minnesota

Galen McKinley
Atmospheric and Oceanic Services
University of Wisconsin-Madison

Introduction

North America's inland ocean, the Great Lakes (Figure 7.1), contains about 23,000 km^3 (5,500 cu. mi.) of water (enough to flood the continental United States to a depth of nearly 3 m), and covers a total area of 244,000 km^2 (94,000 sq. mi.) with 16,000 km of coastline. The Great Lakes comprise the largest system of fresh, surface water lakes on earth, containing roughly 18% of the world supply of surface freshwater. Reservoirs of dissolved carbon and rates of carbon cycling in the lakes are comparable to observations in the marine coastal oceans (e.g., Biddanda et al. 2001) (Table 7.1). The drainage area of the Laurentian system (including the Saint Lawrence River) is approximately 1.0 million km^2—approximately one-third of the Mississippi River watershed or roughly 4% of the surface area of North America. The Great Lakes drain through the Saint Lawrence River, which flows approximately 1200 km before emptying into the largest estuary in the world, the Gulf of Saint Lawrence. This feature of the Great Lakes system is unique in relation to the other marginal regions: exchange with the open ocean is in only one direction, to the ocean. Because of the large size of the watershed, physical characteristics such as climate, soils, and topography vary across the basin. Terrestrial and atmospheric forcing is strongly latitude-dependent in this large basin. To the north, the climate is cold and the terrain is dominated by a granitic bedrock called the Canadian (or Laurentian) Shield consisting of Precambrian rocks under a generally thin layer of acidic soils. Conifers dominate the northern forests. In the southern areas of the basin, the climate is much warmer. The soils are deeper with layers or mixtures of clays, carbonates, silts, sands, gravels, and boulders deposited as glacial drift or as glacial lake and river sediments. The lands are usually fertile and have been extensively drained for agriculture. The original deciduous forests have given way to agriculture and sprawling urban development. This variability has strong impacts on the characteristics of each lake. The lakes are known to have significant effects on air masses as they move in prevailing directions, as exemplified by the 'lake effect snow' that falls heavily in winter on communities situated on the eastern edges of lakes. If the Lakes can frequently experience the degrees of CO_2 undersaturation shown below, then the CO_2 of the airmasses must be impacted as well.

Subregions of the Laurentian System

Although part of a single system, each lake is different (Table 7.1), and the system can be classified into three broad sub-categories: Lake Superior, an oligotrophic lake with low anthropogenic impact; the remaining four lakes, which are more productive and extensively anthropogenically impacted; and the Saint Lawrence River. In volume, Lake Superior is the largest, deepest, and coldest of the lakes. Because of its size, Superior has a retention time of 191 years based on the volume of water in the lake and the mean rate of outflow. Most of the forested, granitic Superior Basin is sparsely populated, with little agriculture because of a cool climate and poor soils.

The other four lakes reside in carbonate basins with deeper, more fertile soils, and are subjected to more extensive human activities. Lake Michigan, the largest of these, spans the upper and lower regions of the Laurentian Basin. The northern part is in the colder, less developed, upper Great Lakes region. It is sparsely populated, except for the Fox River Valley, which drains into Green Bay. This bay has one of the most productive Great Lakes fisheries but receives wastes from the world's largest concentration of pulp and paper mills. The more temperate southern basin of Lake Michigan is among the most urbanized areas in the Great Lakes system, containing the Milwaukee and Chicago metropolitan areas. This region is home to about 8 million people or about one-fifth of the total population of the Great Lakes Basin. Lake Huron, which includes Georgian Bay, is the third largest of the lakes by volume. The Saginaw River Basin is intensively farmed and contains the Flint and Saginaw-Bay City metropolitan areas. Lake Erie is the smallest of the lakes in volume and is exposed to the greatest effects from urbanization and agriculture. Because of the fertile soils surrounding the lake, the area is intensively farmed. The lake receives runoff from the agricultural area of southwestern Ontario and parts of Ohio, Indiana, and Michigan. Seventeen metropolitan areas with populations over 50,000 are located within the Lake Erie Basin. It is the shallowest of the five lakes (average depth is only about 19 m) and therefore warms rapidly in the spring and summer, and frequently freezes over

Figure 7.1. Geographic representations of the Laurentian Great Lakes showing a) topography/bathymetry; b) the size of the lakes relative to their drainage basin; c) the size of the lakes relative to the North American Atlantic coastal ocean; and d) the population distribution in the Laurentian basin.

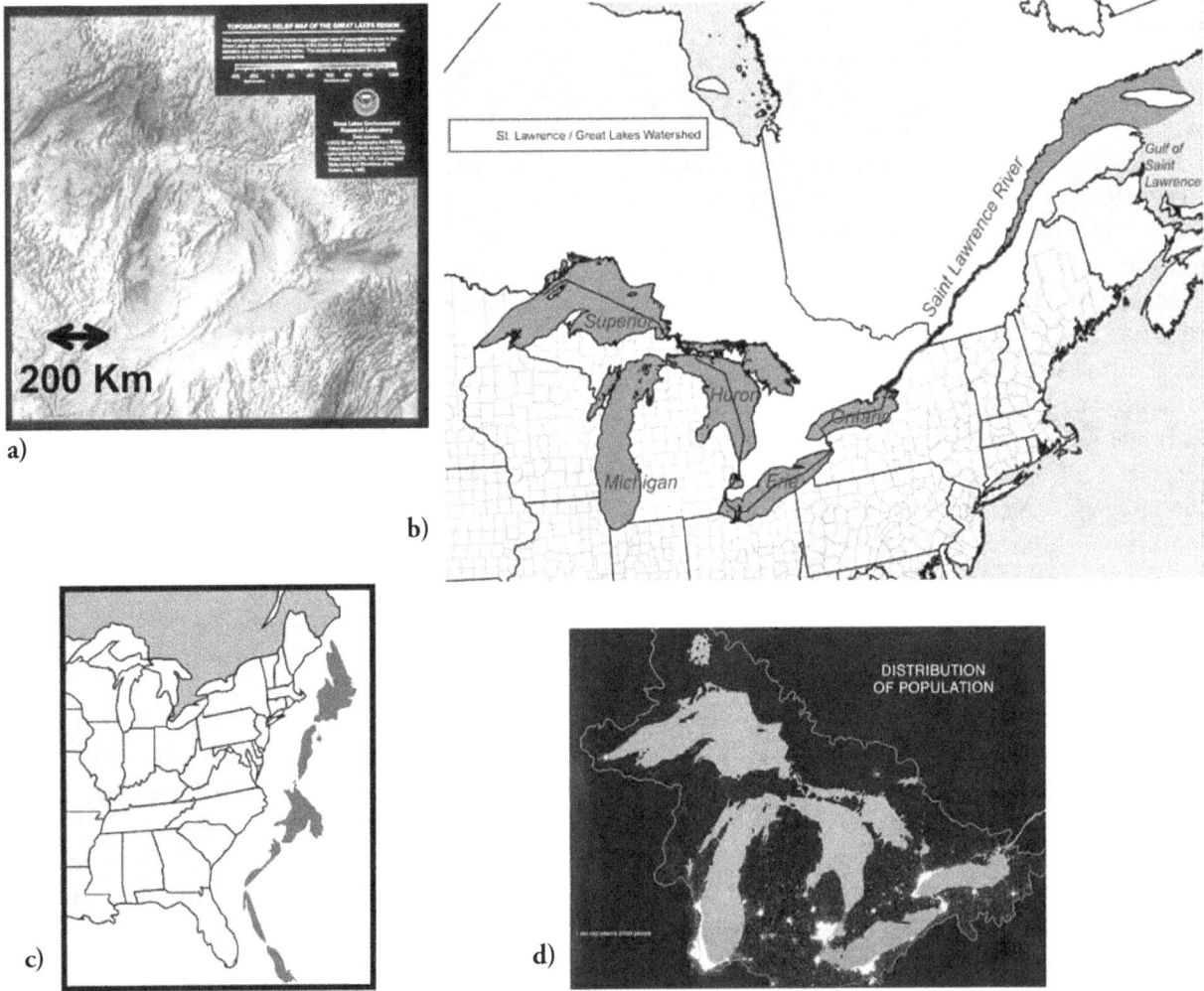

in winter. It also has the shortest retention time of the lakes, 2.6 years. Lake Ontario, although slightly smaller in area, is much deeper than Lake Erie, with an average depth of 86 m (283 ft) and has a retention time of about 6 years. Major urban industrial centers, such as Hamilton and Toronto, are located on its shore. The US shore is less urbanized and is not intensively farmed, except for a narrow band along the lake.

The Saint Lawrence River is born at the outflow of Lake Ontario before draining into the Gulf of Saint Lawrence, the largest estuary in the world. It runs over 3000 km from source to mouth (1,197 km from the outflow of Lake Ontario). Its drainage area covers 1.03 million km². The average discharge at the mouth (into the North Atlantic) is 10,400 m³ s⁻¹.

Table 7.1. Relevant statistics for the Laurentian Great Lakes, and comparative nominal values for the coastal oceans.

Lake	Average Depth (m)	DIC (mM)	DOC (µM)	PP (mmol m⁻² d⁻¹)	Sediment C burial (mol m⁻² yr⁻¹)
Superior	147	1	100		0.25
Michigan	85	2.3	400	12	8e11 g C/y
Huron	59	1.7	400		
Erie	19	2.2	400		2
Ontario	86	2.2	400		2e11 g C/yr
Coastal Oceans	≈100	≈2	≈100		

Carbon Cycling in the Laurentian System

Carbon cycling is quite different among the lakes. At one extreme (Figure 7.2) is Lake Superior, a unique, ultra-oligotrophic system with many features similar to the oligotrophic oceanic gyres, such as dominance of microbial biomass and dissolved organic carbon (DOC) in biogeochemical processes (see e.g., Biddanda et al. 2001). The combination of cold temperature, low nutrient concentrations (total phosphorus <2 µg L⁻¹), and low municipal/industrial pressure leads to rates of primary production, estimated at 2.0 to 6.7 Tg C yr⁻¹ (Urban et al. 2005), or approximately 6 to 20 mmol C m⁻² d⁻¹, that are among the lowest measured in any aquatic system. Low primary production and the soil-starved granitic drainage basin result in low

Figure 7.2. Schematic of a carbon budget for Lake Superior, the most oligotrophic of the lakes. Reservoirs are expressed in units of Tg C, and fluxes (arrows) in Tg C yr⁻¹ (Cotner et al. 2004).

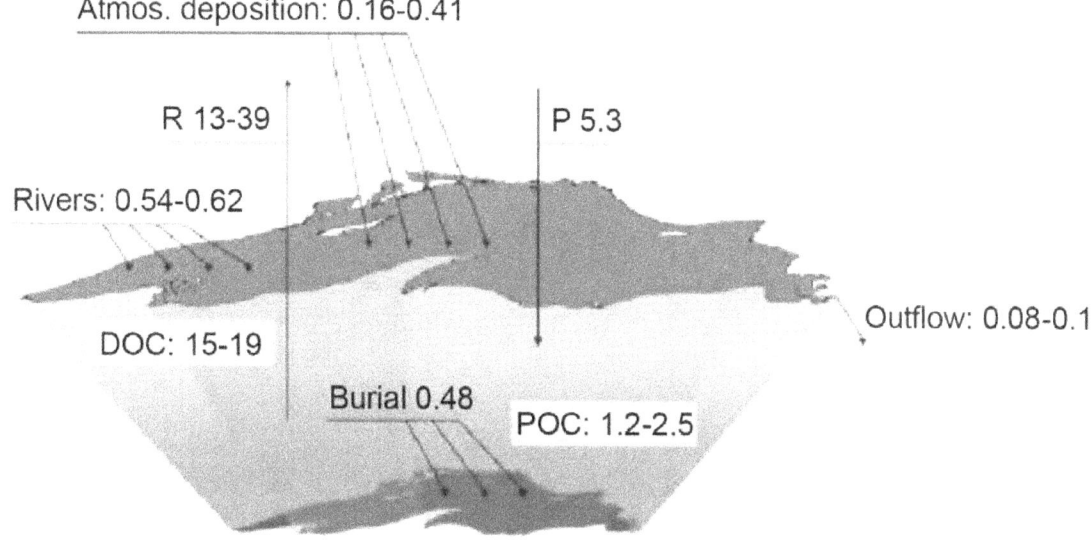

Atmos. deposition: 0.16-0.41

R 13-39   P 5.3

Rivers: 0.54-0.62

DOC: 15-19   Outflow: 0.08-0.1

Burial 0.48   POC: 1.2-2.5

DIC, DOC, and POC concentrations in the lake. Allochthonous riverine organic carbon inputs were estimated at 0.5 to 0.6 Tg C yr^{-1}, which is about 10% of photo-autotrophic production. Atmospheric carbon deposition has not been measured to any significant extent but we estimate it at 0.16 to 0.41 Tg yr^{-1} (Cotner et al. 2004). All together, allochthonous carbon sources provide 13 to 19% of photo-autotrophic production. The main loss of organic matter in the lake is through respiration in the water column at a lake-wide total of 13 Tg C yr^{-1} (estimate range: 13–81 Tg C yr^{-1}; Urban et al. 2005; Cotner et al. 2004). Respiration is (at a minimum) double all estimated organic carbon sources combined and therefore sources are likely underestimated.

The other lakes are quite different. Much warmer than Superior and receiving more substantial carbon and nutrient inputs from the increased anthropogenic activity and deeper soils in their drainage basins, their primary production averages approximately 25 (±8)

mmol C m^{-2} d^{-1}, or higher than the upper-bound estimates seen in Lake Superior. As a result, DIC and DOC concentrations are much higher as well, ranging from 1.7 to 2.3 mmol L^{-1} and 200 to 400 µmol L^{-1}, respectively. Complete carbon budgets for the other lakes have not been rigorously performed, however, the budget for the eutrophic southern Green Bay (Figures 7.3 and 7.4) may serve as an example of these systems.

Net CO_2 Exchange with the Atmosphere

The magnitude of the net air-water CO_2 flux has not been systematically measured in any of the lakes. The lakes are generally agreed to be net heterotrophic, and thus a source of CO_2 to the atmosphere. However, this is based largely on poorly constrained estimates of primary production and respiration. We are aware of only a few direct measurements of pCO_2 in the Great Lakes. These include the efforts of Jim Waples (Great Lakes WATER Institute-UWM) in Green Bay, Lake

Figure 7.3. Schematic of a carbon budget for Green Bay, Lake Michigan that is exemplary of the more eutrophic lakes (from Klump et al. 2007).

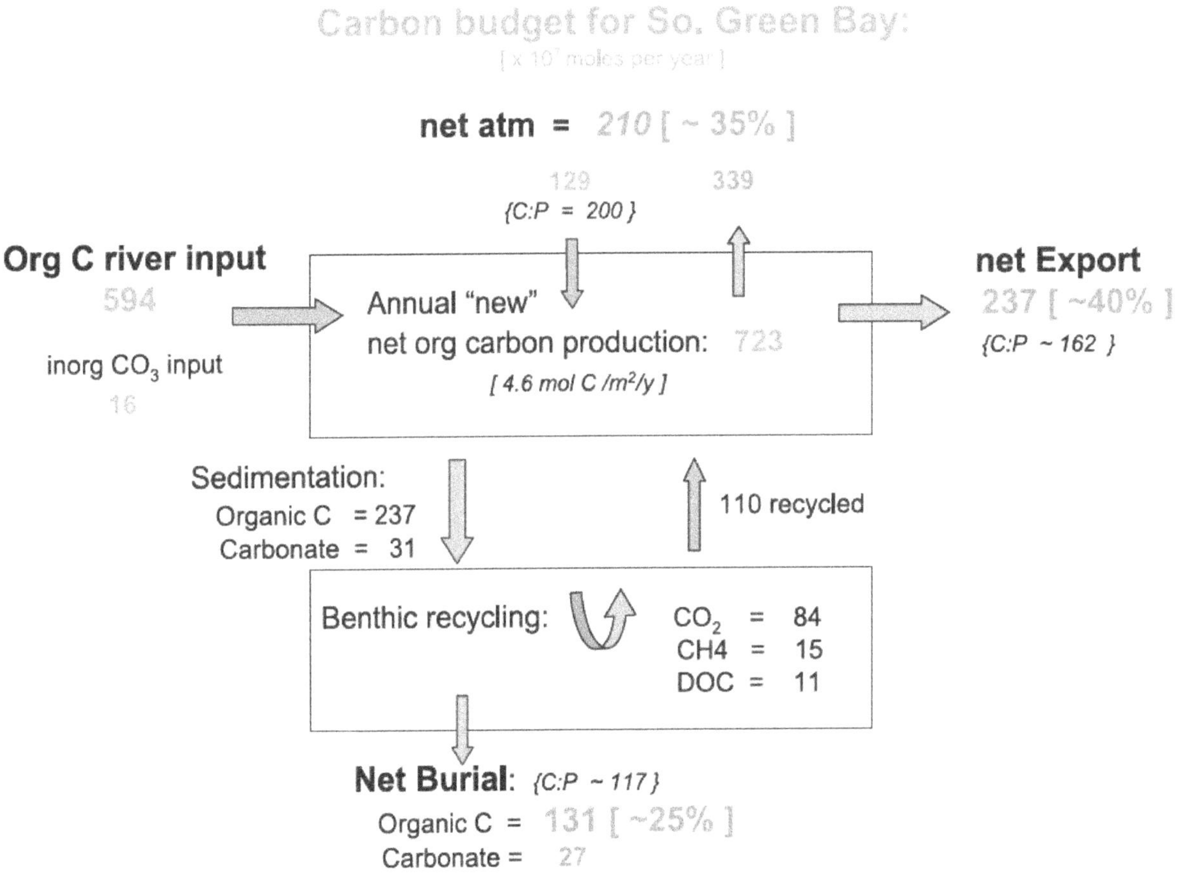

Figure 7.4. Surface $CO_{2(aq)}$ distributions in Green Bay (from Waples, 1998). Net CO_2 efflux (~15 g C m^{-2} yr^{-1}) driven in part by degradation of external inputs of organic C to the system.

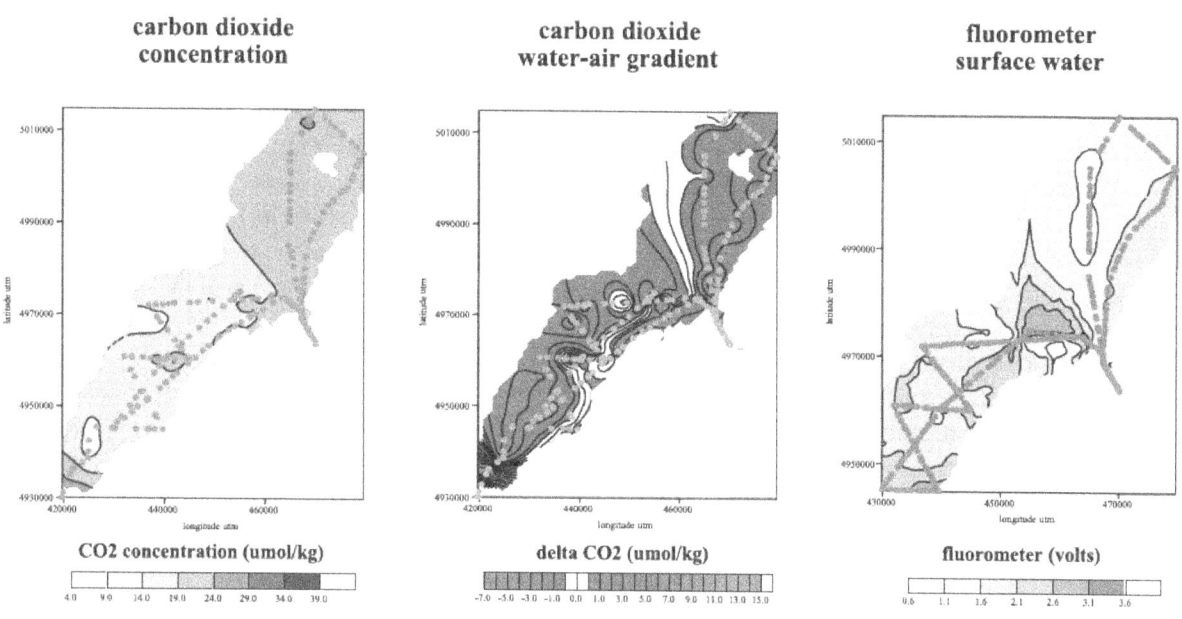

Michigan, and Maggie Squires (University of Waterloo) in Lake Erie. The results of these measurements are equivocal and highly variable. Squires' measurements (Figure 7.5) show surface pCO_2 values as low as 120 µatm, far below atmospheric saturation. Waples' measurements in Green Bay (Figure 7.4) showed a net export of about 0.2 Tg CO_2 to the atmosphere in 1994 and 1995. However, the spatial and temporal direction and magnitude of flux were far from uniform. Riverine loading and annual temperature cycles generate spatial

and temporal gradients in carbon processing, such that isolated measurements in time or space would significantly miss the dynamic nature of the carbon air-sea interactions.

Historical Measurement Programs

There have been no large, coordinated measurement programs focusing on the C cycle in the Great Lakes. This, coupled with the high temporal and spatial

Figure 7.5. Distributions of surface pCO_2 values in Lake Erie. From Squires et al.. (in prep.).

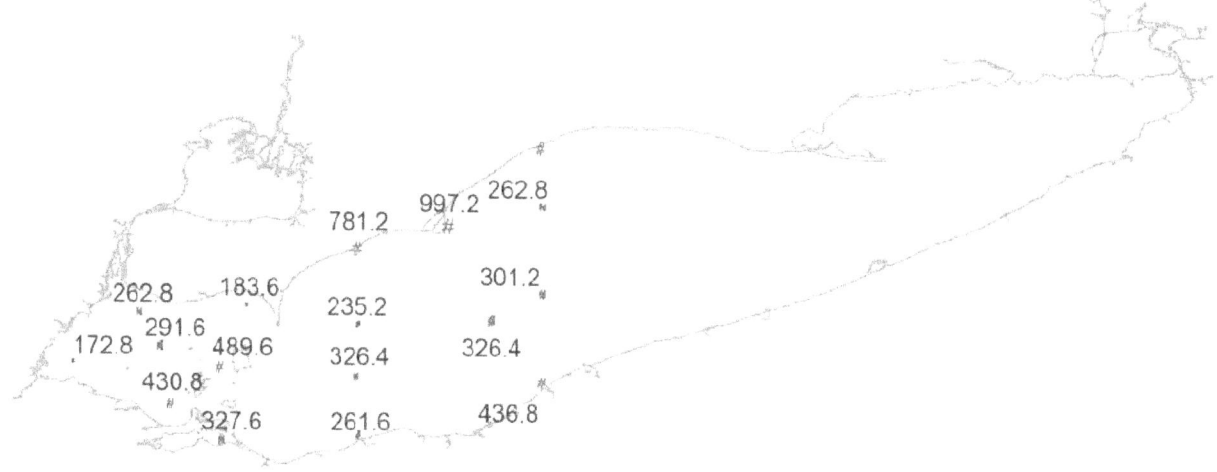

variability in the lakes leaves many key questions regarding net transport and processing of C in the lakes only weakly constrained (Table 7.2).

Anthropogenic impacts on the Great Lakes system

Direct anthropogenic forcing on the Great Lakes is intense, and has probably been a factor in the carbon budget since harvesting of the basin's conifer forests in the late 1800s began to release excess phosphorus, the nutrient limiting primary production in the Great Lakes, to the system. The loads of phosphorus to the lakes due to agricultural and industrial pollution accelerated until the early 1970s, when restrictions were imposed by international treaty (Figure 7.6). Target P loads were established for each lake, based on various ecosystem models, to achieve agreed chlorophyll levels. Because of its small size and high population, Lake Erie was the most impacted. There was a rapid drop in both P and chlorophyll concentrations. However recent (post-1995) concentrations of P in Lake Erie have risen and the cause is not known, although several theories have been proposed (increased P loads, warmer temperatures, and zebra mussel effects). In any case, changes in P cycling imply changes in C cycling, although there have been few C measurements in recent years.

In addition to discharge of pollutants, the carbon cycle in the Great Lakes has been impacted by invasive species. A major pathway for invasive introduction is via international maritime shipping. The lakes were opened to the Atlantic when the St. Lawrence Seaway was completed in the 1950s. Since then there have been several devastating invasions: the most recent in this category are Dressenid mussels (Zebra and Quagga), first discovered in the lakes in 1989. Dreissenid mussels have been found in all of the Great Lakes and have spread from there into the rivers of the Northeast and Midwest. Illustrating the impact that mussels have had on carbon cycling is the approximately 0.1 mmol L^{-1} drop in calcium concentration seen in Lake Erie since the mussels' invasion. This is due to formation of $CaCO_3$ shell material (Figure 7.7), and accompanies an equivalent drop in DIC, and a two-fold greater decrease in alkalinity. This extrapolates to a loss of approximately 1×10^{12} g dissolved C for Lake Erie alone, and translates to significant changes in the carbonate equilibria (and hence the pCO_2 distributions) in the lakes. The mussels may also bring about greater retention and recycling of nutrients in the nearshore (i.e., the 'near shore phosphorus shunt'; Hecky et al. 2004) and in turn increased production in the benthic community. An important question relevant to carbon cycling processes is the magnitude and long-term fate of both

Table 7.2. Knowledge matrix for the carbon cycle in the Great Lakes, where a score of 1 represents a high degree of knowledge about a certain process within a certain lake, and a score of 3 represents a low degree of knowledge.

Process	Superior	Michigan	Huron	Erie	Ontario	Mean "score"
Outflow	1	1	1	1	1	1.0
Hydrodynamic models	2	1	2	1	2	1.6
C Inputs	3	1?	3	1	1	1.8
Sedimentation	3	1	3	2	2	2.2
PP	3	2	3	2	2	2.4
Respiration	3	2	3	2	2	2.4
BGC models	3	3	3	2	3	2.8
CO_2 Exchange	3	3	3	3	3	3.0
Mean "score"	2.6	1.8	2.6	1.8	2.0	

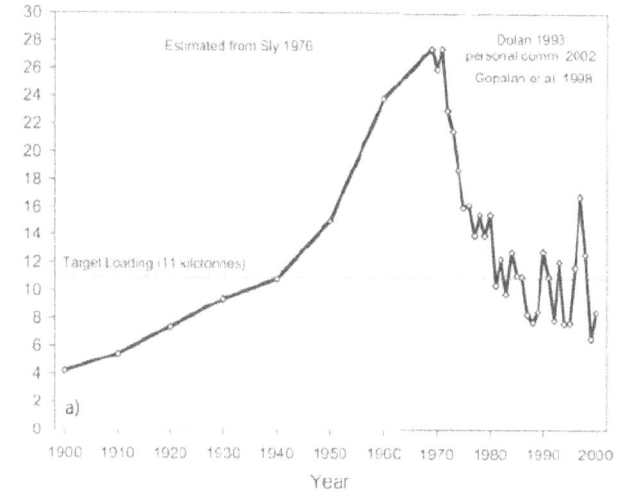

Figure 7.6. The history of phosphorus loading in Lake Erie (a) showing the steady rise in inputs prior to the US-Canadian treaty in 1973, and the sharp drop afterwards. The concentrations of phosphorus in the lake (b) and chlorophyll levels (c) both decreased in response.

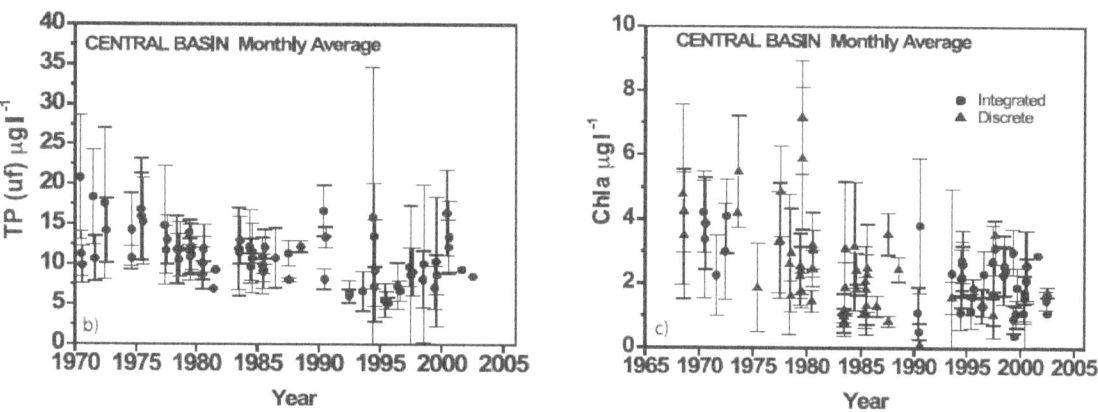

Figure 7.7. The Zebra mussel invasion in the Great Lakes. The large deposits of calcareous shell material (a) led to a significant drop in the alkalinity and (b) calcium content of Lake Erie after the accidental introduction of Dressenid mussels to the lake.

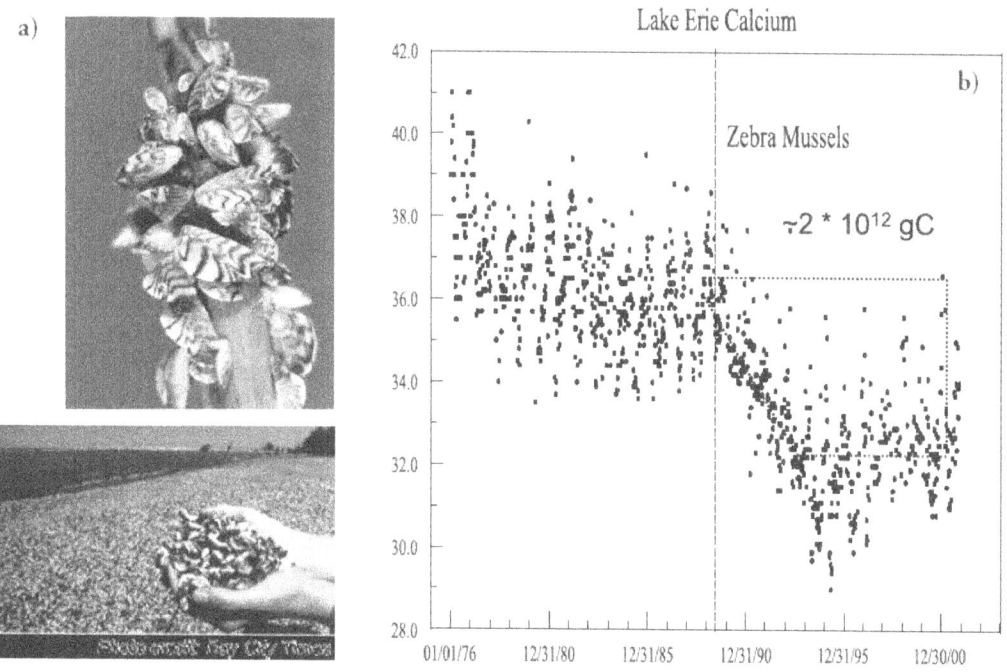

Dreissenid biodeposits and benthic algal biomass. Benthic organic carbon may be transported down-lake and lost at the outflow (Williams et al. 2000), rapidly buried in the nearshore (Howell et al. 1996) or pelagic zone. Evidence suggests that carbon burial rates in the pelagic zone may be decreasing, thereby reducing offshore nutrient availability and respiration rates and phytoplankton production.

The Lakes are also sensitive to large-scale climate change. Nine of the ten warmest years on record in the Great Lakes Basin have occurred since 1990 (Figure 7.8). 1998 was the warmest year, with the mean temperature at the mid point of the regional forecast for 2050. When all data have been collected, 2005 will either replace 1998 or be the second warmest year. The consequences for carbon cycling are complex and not yet fully modeled. Less ice cover, earlier thermal stratification, warmer/thicker upper mixed

layers, and generally faster rates for carbon-related processes all need to be evaluated. Another immediate concern relates to a recently observed shift in the mean summer wind field over the Great Lakes Basin. Waples and Klump (2002) examined hourly, NDBC buoy-recorded, summertime wind vectors spanning a 20-year period beginning in 1980 and found that a major shift in wind direction occurred at—and has persisted since—the end of the 1980s. The most likely explanation for this shift in wind direction is a southward displacement of the dominant summer storm track. Wind-driven changes in estuarine circulation and coastal hydrodynamics have been shown to affect water column profiles of temperature, salinity, and dissolved oxygen (Goodrich et al. 1987; Welsh and Eller, 1991) as well as fish and invertebrate populations (Kilgour et al. 2000; Officer et al. 1984).

Figure 7.8. Temperature history of Lake Michigan. Nine of the warmest years on record occurred within the last ten years, with 1998 and 2005 ranking first and second, respectively. This trend has coincided with a decrease in the duration of ice cover on Lake Michigan.

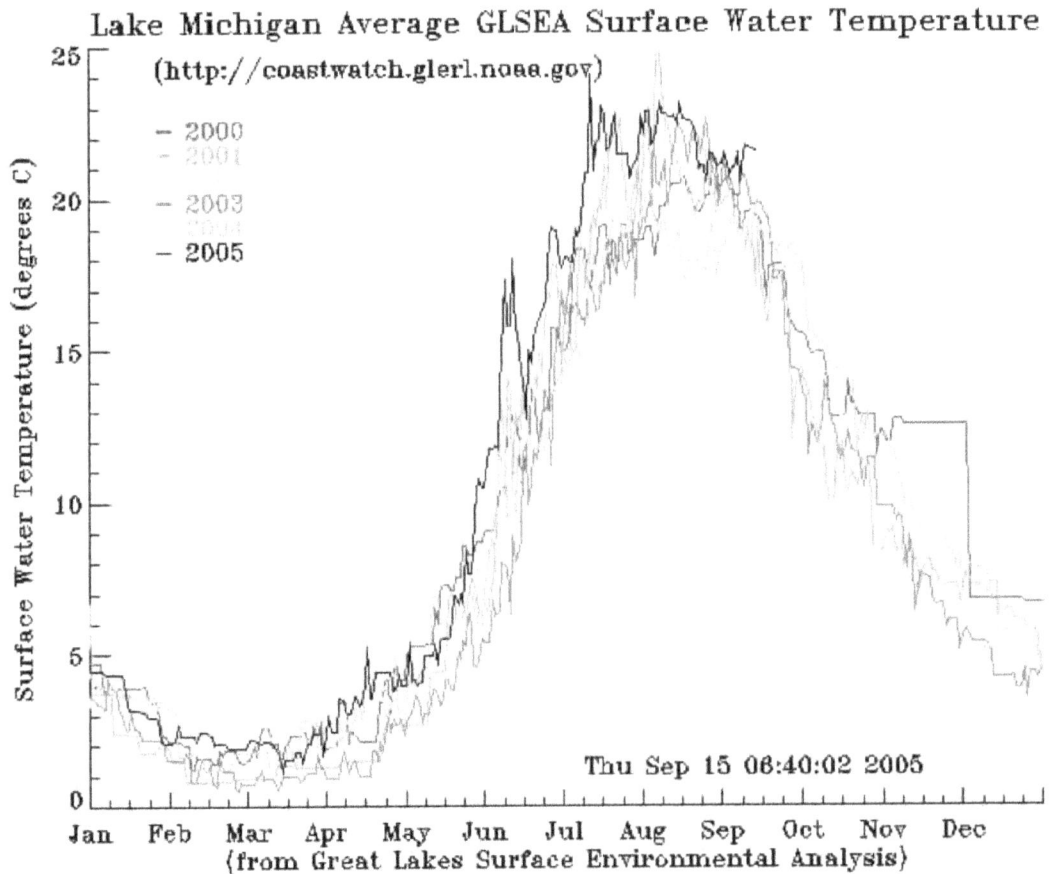

References

Biddanda, B., M. Ogdahl, and J. Cotner, 2001. Dominance of bacterial metabolism in oligotrophic relative to eutrophic waters. *Limnology and Oceanography*, 46:730-739.

Cotner, J.B., B.A. Biddanda, W. Makino, and T. Stets, 2004. Organic carbon biogeochemistry of Lake Superior. *Aquatic Ecosystem Health and Management*, 7:451-464.

Dolan, D.M., 1993. Point-source loadings of phosphorus to Lake Erie - 1986-1990. *Journal of Great Lakes Research*, 19:212-223.

Goodrich, D.M., W.C. Boicourt, P. Hamilton, and D.W. Pritchard, 1987. Wind-induced destratification in Chesapeake Bay. *Journal of Physical Oceanography*, 17:2232-2240.

Gopalan, G., D.A. Culver, L. Wu, and B.K. Trauben, 1998. Effects of recent ecosystem changes on the recruitment of young-of-the-year fish in western Lake Erie. *Canadian Journal of Fisheries and Aquatic Sciences*, 55:2572-2579.

Hecky, R.E., R.E.H. Smith, D.R. Barton, S.J. Guildford, W.D. Taylor, M.N. Charlton and T. Howell. 2004. The nearshore phosphorus shunt: a consequence of ecosystem engineering by dreissenids in the Laurentian Great Lakes. *Canadian Journal of Fisheries and Aquatic Sciences*, 61:1285-1293.

Howell, E.T., C.H. Marvin, R.W. Bilyea, P.B. Kauss, and K. Somers, 1996. Changes in environmental conditions during Dreissena colonization of a monitoring station in eastern Lake Erie. *Journal of Great Lakes Research*, 22:744-756.

Kilgour, B.W., R.C. Bailey, and E.T. Howell, 2000. Factors influencing changes in the nearshore benthic community on the Canadian side of Lake Ontario. *Journal of Great Lakes Research*, 26:272-286.

Klump, J.V., S.A. Fitzgerald, and J.T. Waples, 2007. Benthic biogeochemical cycling, nutrient stoichiometry, and carbon and nitrogen mass balances in a eutrophic freshwater bay. *Limnology and Oceanography* (in revision).

Officer, C.B., R.B. Biggs, J.L. Taft, L.E. Cronin, M.A. Tyler, and W.R. Boynton, 1984. Chesapeake Bay anoxia: Origin, development, and significance. Science, 223:22-27.

Sly, P.G., 1976. Lake Erie and its basin. *Journal of the Fisheries Research Board of Canada*, 33:355-370.

Waples, J.T., 1998. Air-Water Gas Exchange and the Carbon Cycle of Green Bay, Lake Michigan. Ph.D. dissertation, University of Wisconsin-Milwaukee, 429 pp.

Waples, J.T., and J.V. Klump, 2002. Biophysical effects of a decadal shift in summer wind direction over the Laurentian Great Lakes. *Geophysical Research Letters*, 29:1201, doi:10.1029/2001GL014564.

Welsh, B.L., and F.C. Eller, 1991. Mechanisms controlling summertime oxygen depletion in western long-island sound. *Estuaries*, 14:265-278.

Williams, D.J., M.A.T. Neilson, J. Merriman, S. L'Italien, S. Painter, K. Kuntz, and A.H. Kl-Shaarawi, 2000. *The Niagara Rivers Upstream-Downstream Program, 1986/87 - 1996/97*. Report No. EHD/ECB-OR/00-01. Ecosystem Health Division, Environment Canada, Burlington, Ont.

Urban, N.R., M.T. Auer, S.A. Green, X. Lu, D.S. Apul, K.D. Powell, and L. Bub, 2005. Carbon cycling in Lake Superior. Journal of Geophysical Research, 110:C06S90, doi:10.1029/2003JC002230.

North American Rivers and Estuaries

James Bauer
School of Marine Science/VIMS
College of William and Mary

Miguel Goni
College of Oceanic and Atmospheric Sciences
Oregon State University

Brent McKee
Department of Marine Sciences
The University of North Carolina at Chapel Hill

Introduction

Rivers are the primary active interface between land and oceans, connecting over 87% of earth's land surface area to the coastlines (Ludwig and Probst, 1998). Ten of the world's largest 50 rivers (by annual discharge) drain the North American continent, primarily along the northern east (Saint Lawrence) and west (Sacramento, Columbia, Fraser) coasts, the Arctic coast (Yukon, Mackenzie, Nelson, Koksauk/Caniapiscau) and northern (Mississippi) and southern (Usumacinta) Gulf coasts (Dai and Trenberth, 2002; Figure 8.1). Many other smaller rivers distributed throughout the North American coasts also make a significant contribution in composite (Dai and Trenberth, 2002). Rivers interact with the open ocean through their estuaries, regions of transition from predominantly freshwater to predominantly ocean water. Some of these estuaries exist in semi-enclosed basins landward of the coastline, while some extend seaward nearly to the shelf break. River discharge is directly forced by atmospheric processes, such as precipitation and evaporation, and by retention and release of moisture in snow-pack, in the forms of surface and subsurface flows.

River plume and estuarine systems are spatially limited, with lateral dimensions of 1 to 100 km and relatively shallow depths of up to tens of meters, yet span maximal ranges in fundamental chemical and physical conditions like salinity and turbidity. Likewise, extreme differences in concentrations of bioreactive species (carbon, nutrients, trace elements) exist over some fraction of the physical dimensions of the estuary/plume (e.g., Hobbie, 2000; Bianchi, 2007). The spatial scales of variability thus range from meters to kilometers. Temporal variability in these systems covers a wide range of scales, given their sensitivity to tidal forcing (time scales of hours); to flood and storm events (days); to seasonal variability in insolation, precipitation, and snowmelt (months); and to long-term climatic variability (e.g., ENSO, PDO) that drives changes in precipitation, winds, and temperatures (years to decades).

Classification of North American Rivers and Estuaries

Sub-categorization of North American rivers and estuaries is not as straightforward as was the case for the other coastal regions described in previous chapters

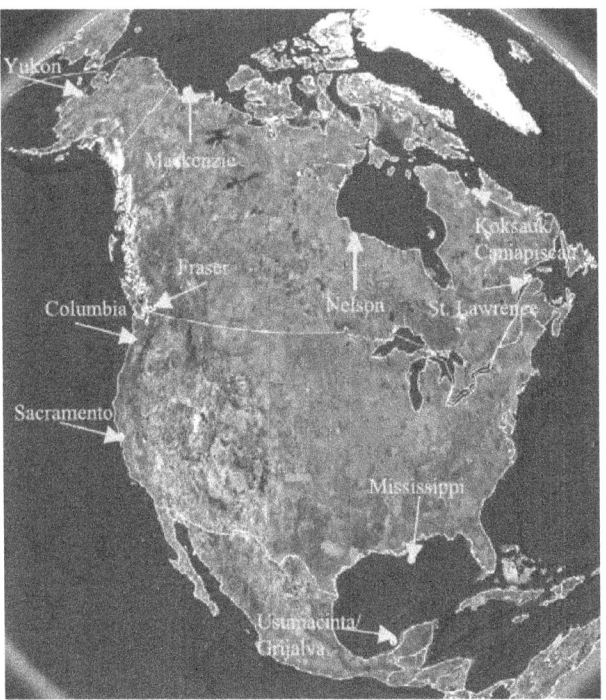

Figure 8.1. Map showing the locations of major North American river systems' points of delivery into the coastal oceans.

of this report because in many ways each river/estuary pair is unique. Further, a given river/estuary system may often be described by a number of different classifications, some of which will vary as hydrographic and oceanographic conditions change over different time scales (e.g., Jay et al. 2000; Vorosmarty and Peterson, 2000; Bianchi, 2007). For example, systems might be distinguished by the hydrographic forcing of their discharge curves, with those dominated by seasonal melting of snow pack distinguished from those dominated by the immediate impacts of local precipitation, or from those dominated by groundwater inputs. Alternatively, systems might be classified by the residence times of river waters within their enclosed estuaries, with the low-discharge, large estuarine systems of the Atlantic coast delineated from the high-discharge, small-estuary rivers draining to the Arctic and the northern Pacific. Still another classification might be based on watershed characteristics, such as total discharge per unit area of drainage basin, or the elevation change between headwaters and estuary. Finally, simple macroscopic features such as total discharge or latitude can also be helpful in classifying and ranking systems. Any one of these classifications may have significance with respect to the carbon cycling within an individual system.

Carbon Cycling in River-Estuary Systems

In spite of some uncertainties, it is generally agreed that rivers globally discharge approximately 1 Pg C yr^{-1}, about 40% in the form of dissolved inorganic carbon, with the remainder delivered in approximately equal parts dissolved and particulate organic carbon (e.g., Degens et al. 1991; Ittekkot and Lane, 1991; Spitzy and Ittekkot, 1991; Spitzy and Leenheer, 1991; Hedges and Keil, 1995; Hedges et al. 1997; Meybeck and Vorosmarty, 1999; Aitkenhead and McDowell, 2000; Schlunz and Schneider, 2000). The total organic carbon discharged from North American rivers is a relatively small proportion of the global total, with estimates ranging from ~0.03 (Figure 8.2) to 0.2 Pg C yr^{-1} (Ittekkot, 1988). In

addition, rivers deliver nutrients, from both natural and pollution sources, that can fuel photosynthetic production of organic carbon in both the river and its associated estuary, as well as in the coastal ocean beyond (e.g., Bierman et al. 1994; Hickey and Banas, 2003; Carmack et al. 2004; Davies, 2004; Wysocki et al. 2006). The input of organic matter by rivers to the oceans is a significant term in the global and oceanic carbon budgets (Hedges, 1992; Berner and Berner, 1996; Hedges et al. 1997; Wetzel, 2001) and occurs primarily via estuaries. Hence, the ~0.25 Pg of DOC discharged from rivers annually can account for the mean radiocarbon-based turnover times of oceanic DOC (~4,000-6,000 yr). The similar magnitude of POC discharged annually by rivers exceeds the long-term burial of organic carbon in ocean sediments (~0.1 Pg yr^{-1}, Berner, 1989; Hedges and Keil, 1995).

In spite of these significant inputs, only small amounts of terrestrial organic matter have been

Figure 8.2. The distributions of total organic carbon discharge by North American rivers into their estuaries. Data summarized from Meybeck (1982), Mulholland and Watts (1982), Meybeck (1993), Ludwig et al. (1996), Aitkenhead and McDowell (2000), and references therein.

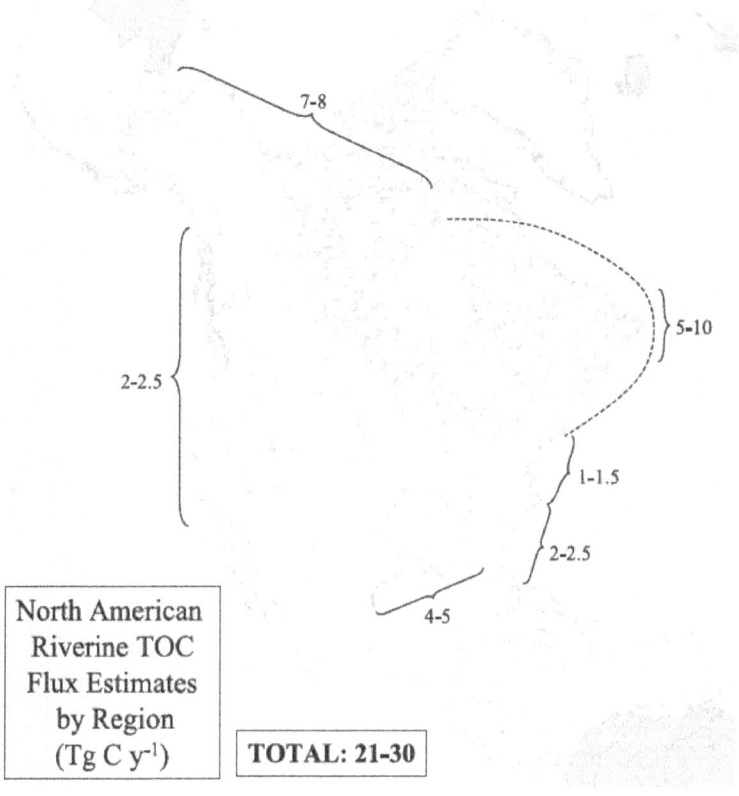

identified in seawater and ocean sediments using organic biomarker and stable carbon isotopic approaches (e.g., Moran et al. 1991; Moran and Hodson, 1994; Benner et al. 1997; Opsahl and Benner, 1997; Hedges and Keil, 1995, Hedges et al. 1997; de Haas et al. 2002; McKee et al. 2004; Stein and Macdonald, 2004). Estuarine processes (chemical, sedimentological, photochemical, and microbial) can significantly alter the molecular composition, isotopic signatures, reactivity, and properties of river-transported terrestrial and autochthonous DOM and POM before it discharges to the coastal zone (e.g., Fox 1991; Moran et al. 2000; Benner and Opsahl, 2001; Minor et al. 2001; Opsahl and Zepp, 2001; Mopper and Kieber, 2002; Aller et al. 2004; Gordon and Goni, 2004; Dagg et al. 2005; Goni et al. 2005a). Past estimates of terrestrial DOM and POM contributions to the oceans have been based on comparisons of the biochemical and isotopic compositions of open ocean and freshwater riverine end-members and have not typically accounted for changes during transit within rivers themselves (e.g., Richey et al. 1990; Cole and Caraco, 2001; Blair et al. 2003; Meybeck and Vorosmarty, 2005), in estuaries (e.g., Raymond and Bauer, 2000,

2001a-c; Benner, 2002; Bianchi et al. 2002; Raymond and Hopkinson, 2003; Goni et al. 2005a; Hoffman and Bronk, 2006), or in ocean margins (e.g., Kieber et al. 1990; Prahl et al. 1994; Bianchi et al. 1997; Bauer and Druffel, 1998; Bauer et al. 2001, 2002; Aller et al. 2004; Gordon and Goni, 2004; Dagg et al. 2005). Thus, without a better understanding of the types and magnitudes of the modifications that components of riverine DOM and POM undergo in estuaries, we may be misinterpreting the "effective" chemical and isotopic signatures (i.e., source specificity) and bioavailability of riverine DOM and POM discharged to the coastal ocean and exported to the ocean interior.

Most notably, virtually all of the discharge estimates for rivers are based on data collected at gauging sites upstream of tidal influences. In large rivers, these sites can be located hundreds of kilometers from the river mouth and do not take into account large regions of lowland rivers and associated floodplains (Figure 8.3). Thus, important questions remain on the exact magnitude and composition of the materials being delivered to the ocean by river/estuary systems (e.g., Fisher et al. 2000). However, it is clear that a large fraction of the fluvial organic material is degraded or altered in estuaries, the coastal ocean, and beyond. Little is known about how processes in the tidally influenced reaches of rivers affect the fluxes and composition of POC, DOC and nutrients (both inorganic and organic), sediment and its associated elements, etc., but these regions are clearly zones of significant modification requiring additional study in order to constrain net land-to-ocean fluxes. Ultimately, carbon cycling in estuaries (Figure 8.4) is a poorly constrained and complicated sum of many processes, each of which may rise and fall in relative importance as tides, discharge conditions, and seasons change.

In spite of decades of estuarine studies on the sources and fates of organic materials in estuaries, we still have only a relatively cursory understanding of the processes that control the reactivity of fluvial DOC and POC in the water column and sediments of these systems (Figures 8.4 and 8.5). In the past few years, we have gained considerable appreciation of the role that physical processes affecting water and particle delivery (e.g., Nittrouer and Wright,

Figure 8.3. Locations of end-member stations relative to estuaries. From McKee (2005).

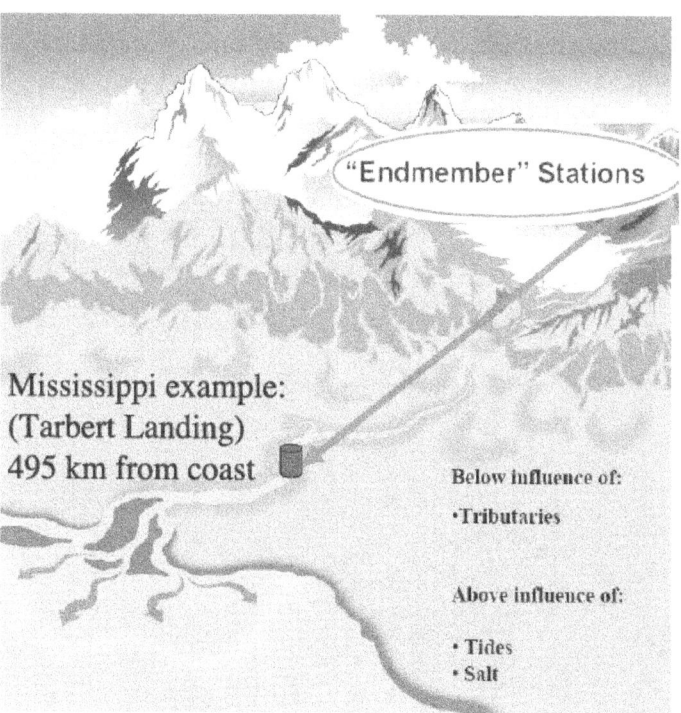

Figure 8.4. General estuarine processes governing the inputs and cycling of carbon and organic matter. Used with permission from Roger Harvey, Chesapeake Biological Laboratory, University of Maryland.

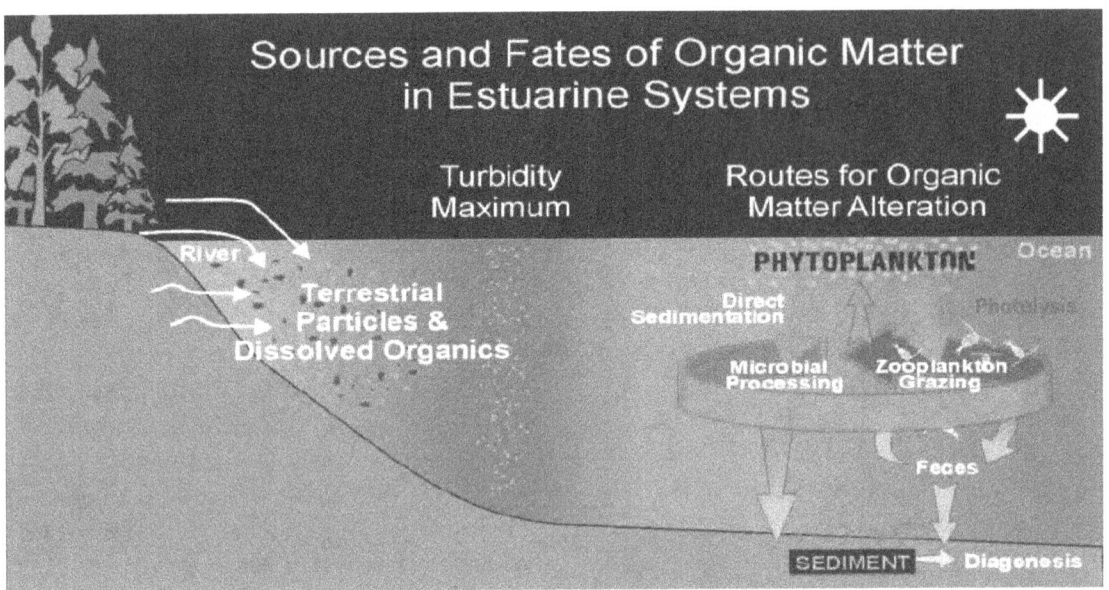

1994; Nittrouer et al. 1996, 2004; Kineke et al. 1996, 2000; Ogston et al. 2000; Wheatcroft and Borgeld, 2000; Geyer et al. 2004; Goni et al. 2005a) play in determining the fate of organic matter throughout the estuary-ocean margin continuum (e.g., Bianchi et al. 1997; Leithold and Hope, 1999; Gordon et al. 2001; Bianchi et al. 2002; Gordon and Goni, 2004; Blair et al. 2004; Dagg et al. 2008). Factors such as the timing of fluvial input in relationship with the dispersal forces acting on the shelf (tides, waves, currents) have critical,

but as yet poorly understood, effects on the efficiency and magnitude of carbon cycling. The connection between physical processes and the biogeochemical processing of organic matter in these systems is a fertile area for future research.

It is increasingly recognized that rivers transport DOC and POC of highly variable age, and, presumably, reactivity, depending upon the specific river and the characteristics of the surrounding watershed. For example, tropical rivers such as the

Figure 8.5. Schematic of POC dynamics in river/estuary/plume settings. Highlighted are some of the components and key processes affecting the dispersal, turnover, and burial of organic matter in these systems. From McKee (2005).

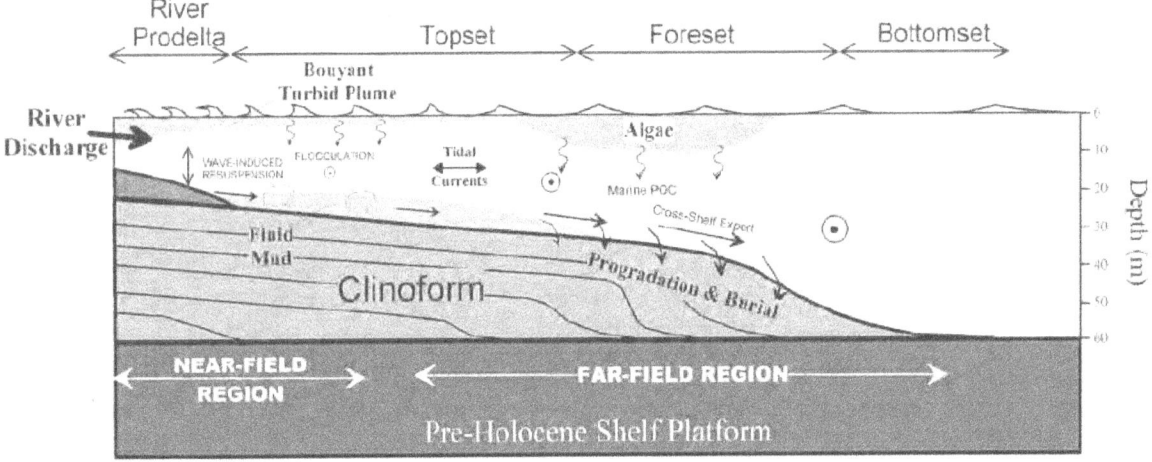

Amazon transport organic materials that are largely modern in nature (i.e., have been fixed on time scales of no more than a few decades), and which are likely influenced by inputs from vegetation in the significant flood plains (Hedges et al. 1986; Mayorga et al. 2005). In contrast, temperate and Arctic rivers have so far been found to contain up to millenial aged POC, and variably aged DOC, ranging from completely modern to several hundred years in age (Raymond and Bauer, 2001a; Raymond et al. 2004; Striegl et al. 2005; Neff et al. 2006).

The nature of the organic matter in a specific river-estuary system is also critically important for long-term sediment preservation. A range of reactivities is generally believed to be present within the organic matter in these systems (Figure 8.6). The conventional wisdom has been that discrete vascular plant debris and freshwater and marine phytoplankton biomass are relatively reactive, while organic material derived from soils and eroded sedimentary rocks is relatively refractory (Prahl et al. 1992; Goni et al. 1998, 2005b, 2006; Masiello and Druffel, 2001; Blair et al. 2003). These latter fractions probably account for the majority of the land-derived organic carbon buried in river-dominated margin sediments, while systems with extensive lowlands and floodplains likely contain more of the former. The relative abundances of these different sources and types of organic materials in the watersheds of different systems, and the correlation between their mobilization as a function of discharge, will likely affect the net carbon burial at any given site (e.g., Blair et al. 2004; Leithold et al. 2005). The complexity of making net flux measurements of different organic matter pools

in settings where both ocean and river waters have varying influence and where high-frequency tidal and wind fluctuations dominate transport is also responsible for the dearth of data (McKee, 2003). In addition, estimates of net respiration in estuarine settings are also lacking, and there is a generally poor understanding of the relative reactivity of terrestrial organic matter in different river, estuarine, and coastal ocean systems.

Air-Water CO_2 Exchange in River/Estuary Systems

Given the uncertainties in the carbon cycle in estuary and river plume systems, it is not surprising that the net effect on the magnitude of the air-water exchange of key carbon-containing gases is also equivocal. Rivers themselves have high pCO_2 (Cole and Caraco, 2001; Figure 8.7), predominantly resulting in net river-to-atmosphere fluxes of CO_2 globally. This, and the potential for significant respiration of riverine and terrestrial organic matter in estuaries suggests that estuaries are also generally sources of CO_2 to the atmosphere (Cai and Wang, 1998; Frankignoulle et al. 1998). Indeed, Borges (2005) suggests that estuaries and salt marshes represent a significant global source of CO_2 to the atmosphere of nearly 0.5 Pg yr^{-1}, which is of the same order of magnitude as that for riverine carbon export that cannot be accounted for in deltaic and shelf sediment deposits. This conclusion is based on a relatively sparse data set, however, and it is unclear how much of this off-gassing is the result of degradation within rivers of terrestrial and river-derived organic material vs. due to the typically high amounts

Figure 8.6. Ranges in the reactivity of fluvial-source organic material for four North American rivers. From Hopkinson et al.. (1998).

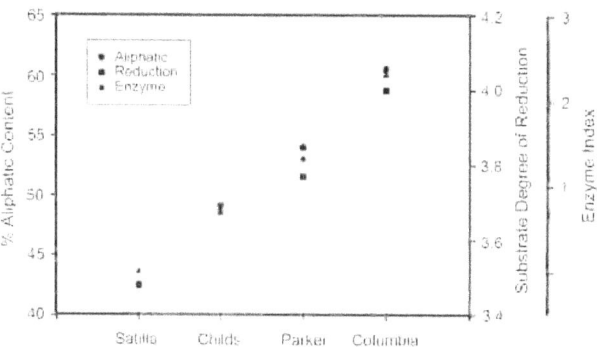

Figure 8.7. pCO_2 levels in world rivers. From Richey et al.. (2002).

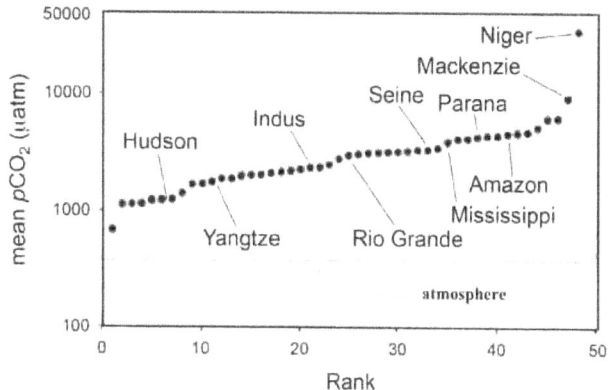

of pCO_2 in river water resulting from a combination of surface runoff and soil respiration (Cole and Caraco, 2001; Richey et al. 2002). Alternatively, elevated productivity and mixing of river and ocean water in estuaries and river plumes can drive pCO_2 below atmospheric saturation (Ternon et al. 2000; Lohrenz and Cai 2006), suggesting that some river/estuary systems can act as sinks for atmospheric CO_2.

Anthropogenic Impacts on River/Estuarine Systems

River/estuary systems have been extensively and directly altered by human activities, and this is only expected to increase in the future as the human population increases further along coastlines and rivers (Postel et al. 1996; Serageldin, 1995; Meybeck and Vorosmarty, 2005). Human activities can have a variety of effects on river and estuarine processes and cycles, sometimes in opposite senses. Deforestation, tillage, and irrigation-enhanced erosion have increased loads of sediments and particulate carbon to rivers

(Ver et al. 1999), while widespread construction of dams, in contrast, has resulted in significant retention of carbon and sediments in reservoirs (Meade et al. 1990). Furthermore, the reduction of river suspended loads by dams around the world has generally resulted in increasing light availability and the potential role of phytoplankton biomass in river and estuarine biogeochemistry (Thorp and Delong, 1994; Humborg, 1997; Humborg et al. 2000; Ittekkot et al. 2000; Sullivan et al. 2001). In the Mississippi watershed, increases in the river's discharge over the last several decades have been found to correspond with increases in the export of carbonate alkalinity; these alkalinity increases further correspond in their magnitude to changing land use (Figure 8.8; Raymond and Cole, 2003). Inputs of nutrients to river/estuarine systems by industrial, municipal, and agricultural runoff have shown significant anthropogenically-driven increases (Howarth et al. 1996; Jickells, 1998; Socolow, 1999), potentially leading to net atmospheric CO_2 uptake in river-influenced margins by increasing primary

Figure 8.8. Changes in Mississippi River alkalinity discharge. From Raymond and Cole (2003).

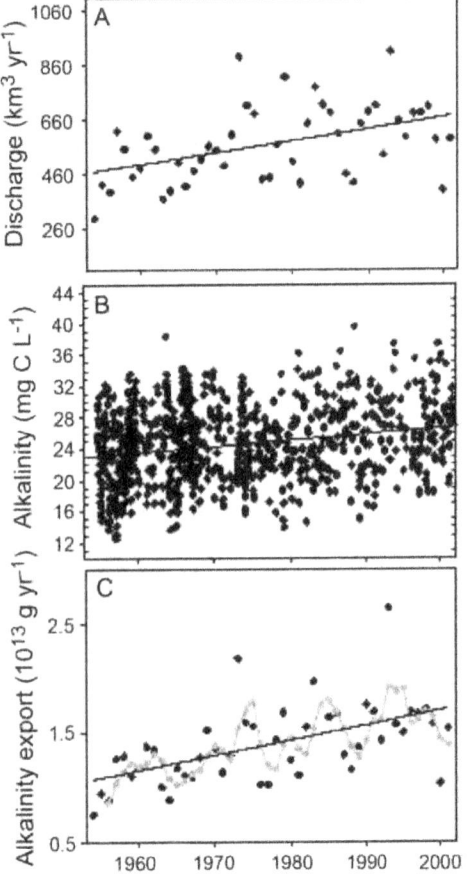

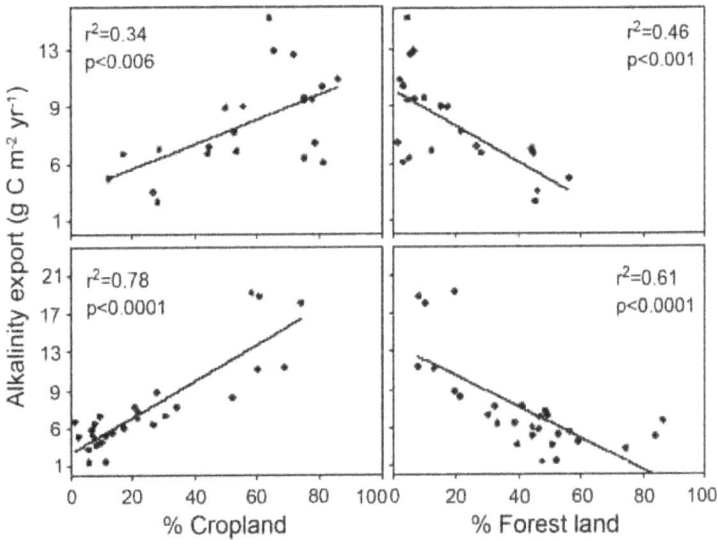

production relative to respiration. Dams may moderate this effect to some extent: for example, P may be retained in reservoir sediments (Jickells, 1998), Si may be retained by decreased weathering rates (Humborg et al. 2000), and nitrate and Si export may be reduced by freshwater diatom blooms in stratified impoundments (Sullivan et al. 2001; Humborg et al. 2000; Pocklington and Tan, 1987). Other human impacts on the geomorphology of the estuary-ocean interface (e.g., dredging, diking, loss of marshes) may have significant effects on sediment transport and carbon burial, and on water residence time and reactions within estuaries.

Estuaries are also tremendously sensitive to long-term global climate change. Changes in river discharge in a warming climate are predicted to be large as snow pack and glacial water retention decreases and precipitation increases. Increased atmospheric CO_2 and warmer temperatures, combined with changes in river discharge to estuaries, could have significant impact on carbon delivery to the oceans (Figure 8.9; Peterson et al. 2002; Frey and Smith, 2005), thus impacting the DOC, alkalinity, and dissolved inorganic carbon (DIC) contents of the primary river water supplied to estuaries. Rising sea levels that outpace the growth of estuary mouth sills (or bars) could subject previously protected estuaries to increased wave-energy regimes, altering the depositional and redox environment in estuarine sediments.

Measurement Programs

Historically, long-term measurement of carbon discharge by rivers has not been the focus of most gauging and monitoring programs. For example, most of the biogeochemical flux measurements in rivers performed by the U.S. Geological Survey are limited to water discharge. Measurements of carbon and nutrient loads needed to calculate fluxes have been carried out only at certain stations on selected rivers, and only for relatively short periods of time as part of targeted studies. Furthermore, as a result of logistical constraints and sampling limitations, particle load and solute measurements are not typically performed during periods of high discharge (i.e., floods), when most material transport and flux tends to occur. In addition, the vast majority of river-based measurements and monitoring efforts are conducted well above the head

of tide. As a result, there is an even larger dearth of information on distributions and fluxes of carbon and other species through the tidal reaches of rivers, and in estuaries and deltaic systems.

Carbon-based measurements have in the past been carried out as parts of specific research projects with generally limited funding lives. These projects have varied from large multi-disciplinary programs, such as the NSF-funded CAMREX program that focused on the biogeochemistry of the Amazon River above tidal influence, to single-investigator efforts focusing on specific aspects the carbon cycle of rivers, estuaries, and continental margins. Programs such as NSF's Integrated Carbon Cycle Research Program and the North American Carbon Program were conceived as possible vehicles for integrated research efforts. Unfortunately, the level of funding support and the wide-open nature of these programs in terms of sites and processes, has made it difficult to carry out well-integrated multi-disciplinary studies. The investigations supported under the auspices of NSF's Coastal Ocean Processes (CoOP) Buoyancy-Driven Transport Processes are an example of multi-disciplinary studies focused on specific sites. However, carbon is not a focus of these studies and there are many compartments of the river-estuarine system (e.g., tidal river, intertidal, and subaqueous sediments) that are not examined as part of the two projects currently supported under this CoOP initiative. The recent NSF Water and Carbon in the Earth System competition holds the promise of more fully integrated studies by supporting interdisciplinary projects that will examine carbon cycling and transfer across the river/estuarine/coastal ocean continuum.

Several large non-carbon based programs that have been supported in the past may be viewed as possible models for future initiatives. Some of these include AMASSEDS, the ONR-funded STRATAFORM (http://faculty.washington.edu/ogston/strataform.shtml), and the NSF MARGINS-sponsored Source-to-Sink (http://www.nsf-margins.org/) programs. In each of these programs, sediment transport and deposition was at the center of the multidisciplinary efforts, which included comprehensive field programs, continuous remote observations, and integrated modeling efforts. The source-to-sink model, which seeks to track the magnitude and composition of river sediments from their generation in uplands to their final sink in ocean margins, is one such appropriate approach by which to study carbon transport and processing along the river/ estuary/ocean continuum. Many of the merits of such a study were exhaustively discussed and clearly identified in the RIOMAR workshops, the results of which are available in public documents (e.g., McKee, 2003; http://www.tulane.edu/~riomar/).

Other relevant ongoing efforts to examine carbon and elemental losses from land and transport by rivers include the EPA Great Rivers project (http://www.epa.gov/emap/greatriver, part of the Environmental Monitoring and Assessment Program (EMAP). This program is designed to establish methods and approaches for assessing and monitoring the ecological condition of the Missouri, Upper Mississippi, and Ohio Rivers in the central United States. Field crews from cooperating state and federal agencies have been sampling organisms, water, and sediments in

the these rivers, with the ultimate goal of developing 'report cards' on the ecological health of each of the sub-systems. The Consortium of Universities for the Advancement of Hydrologic Science (CUAHSI; http://www.cuahsi.org/), an organization representing more than 100 US universities, is supported by the National Science Foundation. The most tangible product of this consortium to date is the Hydrologic Information System (HIS; http://www.cuahsi.org/his.html), an effort to centrally store and publicize water-resource information. The HIS will facilitate data discovery by using a map-based viewer displaying the locations where data has been collected by various entities, including both government- and university-collected data. Data delivery in the HIS platform also allows users to retrieve data directly into databases and spreadsheets and provides analysis packages regardless

Figure 8.9. Potential increases in DOC and river discharge to the Arctic in a warming climate. From Frey and Smith (2005) and Peterson et al. (2002).

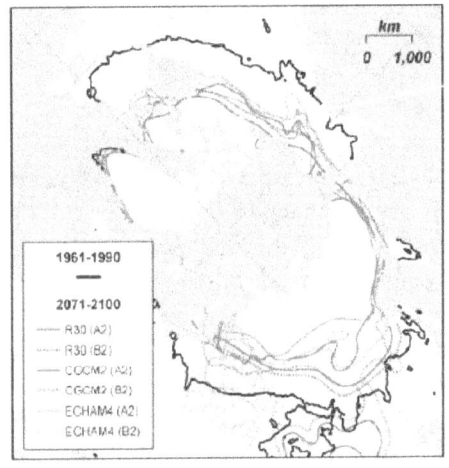

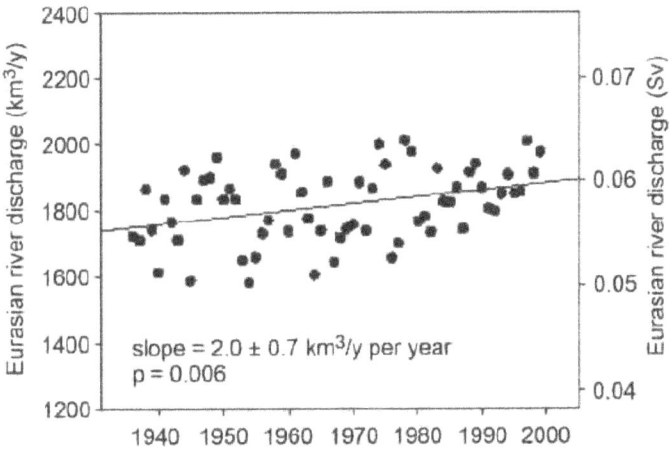

of data source. HIS includes data publication, allowing academic and research scientists to publish data they collected within the common data viewer and responds to the same data retrieval calls as government sources. Data curatorship is provided as part of the HIS Data Center (HISDAC) for the storage and archiving of data.

It should be noted that integrated carbon-cycling approaches and the linkages between head-of-tide and coastal ocean have been largely ignored in studies of North American river and estuary systems. Future programs should focus on measuring net fluxes of all carbon species (e.g., dissolved and particulate inorganic carbon, DOC, and POC) across key interfaces (soil/water, sediment/water, air/water) of contiguous sub-systems (e.g., upland river, tidal river, floodplain, estuary, delta, shelf), coupling these measurements with investigations of the processes responsible for the transformation of carbon in these various reservoirs. Modeling of carbon transformations and fluxes in these systems needs to be carefully integrated with the measurement efforts to ensure seamless and continuous communication between researchers before, during, and after the field efforts. Observatory technology should be incorporated into such programs to provide the necessary continuous-data framework in which

to integrate intensive sampling and modeling efforts. Observatories are also likely to provide the major means for capturing low-frequency events such as floods and storms, which play a key role in the carbon biogeochemistry of these systems. However, observatories alone cannot capture the processes during such stochastic events. Hence, support for rapid-response efforts should be made available to comprehensively study the effects of these critical phenomena. Future programs should also focus on sites where the linkage between systems can be accomplished within the time frame of the funding cycle. Finally, opportunities to mine existing data sets should be made available to researchers in order to take advantage of the wealth of information already collected by previous and on-going monitoring efforts.

References

Aitkenhead, J.A., and W.H. McDowell, 2000. Soil C/N as a predictor of annual riverine DOC flux at local and global scales. *Global Biogeochemical Cycles*, 14:127-138.

Aller, R.C., A. Hannides, C. Heilbrun, and C. Panzeca, 2004, Coupling of early diagenetic processes and sedimentary dynamics in tropical shelf environments: the Gulf of Papua deltaic complex. *Continental Shelf Research*, 24:2455-2486.

Bauer, J.E., and E.R.M. Druffel, 1998. Ocean margins as a significant source of organic matter to the deep open ocean. *Nature*, 392:482-485.

Bauer, J.E., E.R.M. Druffel, D.M. Wolgast, and S. Griffin, 2001. Cycling of dissolved and particulate organic radiocarbon in the northwest Atlantic continental margin. *Global Biogeochemical Cycles*, 15:615-636.

Bauer, J.E., E.R.M. Druffel, D.M. Wolgast, and S. Griffin, 2002. Temporal and spatial variability in sources and cycling of DOC and POC in the northwest Atlantic continental margin. *Deep Sea Research II*, 49:4387-4419.

Benner, R., 2002. Chemical composition and reactivity. In: *Biogeochemistry of Marine Dissolved Organic Matter* [D.A. Hansell and C.A. Carlson (eds.)]. Academic Press, San Diego.

Benner, R., B. Biddanda, B. Black, and M. McCarthy, 1997. Abundance, size distribution, and stable carbon and nitrogen isotopic compositions of marine organic matter isolated by tangential-flow ultrafiltration. *Marine Chemistry*, 57:243-263.

Benner, R., and S. Opsahl, 2001. Molecular indicators of the sources and transformations of dissolved organic matter in the Mississippi river plume. *Organic Geochemistry*, 32:597-611.

Berner, E.K., and R.A. Berner, 1996. *Global Environment: Water, Air and Geochemical Cycles*. Prentice-Hall, Upper Saddle River, NJ, 376 pp.

Berner, R.A., 1989. Biogeochemical cycles of carbon and sulfur and their effect on atmospheric oxygen over Phanerozoic time. *Palaeogeography, Palaeoclimatology, Palaeoecology*, 75, 97-122.

Bianchi, T.S., 2007. *Biogeochemistry of Estuaries*. Oxford University Press, 702 pp.

Bianchi, T.S., C.D. Lambert, P.H. Santschi, and L. Guo, 1997. Sources and transport of land-derived particulate and dissolved organic matter in the Gulf of Mexico (Texas shelf/slope): the use of lignin-phenols and loliolides as biomarkers. *Organic Geochemistry*, 27:65-78.

Bianchi, T.S., S. Mitra, and M. McKee, 2002. Sources of terrestrially-derived carbon in the Lower Mississippi River and Louisiana shelf: Implications for differential sedimentation and transport at the coastal margin. *Marine Chemistry*, 77:211-223.

Bierman, V.J.J., S.C. Hinz, D.-W. Zhu, D.J. Wiseman, N.N. Rabalais, and R.E. Turner, 1994. A preliminary mass balance model of primary productivity and dissolved oxygen in the Mississippi River Plume/Inner Gulf Shelf region. *Estuaries*, 17:886-899.

Blair, N.E., E.L. Leithold, S.T. Ford, K.A. Peeler, J.C. Holmes, and D.W. Perkey, 2003. The persistence of memory: the fate of ancient sedimentary organic carbon in a modern sedimentary system. *Geochimica et Cosmochimica Acta*, 67:63-73.

Blair, N.E., E.L. Leithold, and R.C. Aller, 2004. From bedrock to burial: the evolution of particulate organic carbon across coupled watershed-continental margin systems. *Marine Chemistry*, 92:141-156.

Borges, A.V., 2005. Do we have enough pieces of the jigsaw to integrate CO_2 fluxes in the coastal ocean? *Estuaries*, 28:3-27.

Cai, W.-J., and Y. Wang, 1998. The chemistry, fluxes, and sources of carbon dioxide in the estuarine waters of the Satilla and Altamaha Rivers, Georgia. *Limnology and Oceanography*, 43:657-668.

Carmack, E., R. Macdonald, and S. Jasper, 2004. Phytoplankton productivity on the Canadian Shelf of the Beaufort Sea. *Marine Ecology Progress Series*, 277:37-50.

Cole, J.J., and N. Caraco, 2001. Carbon in catchments: connecting terrestrial carbon losses with aquatic metabolism. *Marine and Freshwater Research*, 52:101-110.

Dagg, M.J., T.S. Bianchi, G.A. Breed, W.-J. Cai, S. Duan, H. Liu, B.A. Mckee, R.T. Powell, and C.M. Stewart, 2005. Biogeochemical characteristics of the Lower Mississippi River, USA, during June 2003. *Estuaries*, 28:664-674.

Dagg, M.J., T.S. Bianchi, B.A. McKee, and R.T. Powell, 2008. Fates of dissolved and particulate materials from the Mississippi River immediately after discharge into the northern Gulf of Mexico, USA, during a period of low wind-stress. *Continental Shelf Research* (in press).

Dai, A., and K.E. Trenberth, 2002. Estimates of freshwater discharge from continents: latitudinal and seasonal variations. *Journal of Hydrometeorology*, 3:660-687.

Davies, P., 2004. Nutrient processes and chlorophyll in the estuaries and plume of the Gulf of Papua. *Continental Shelf Research*, 24:2317-2341.

Degens, E.T., S. Kempe, and J.E. Richey (eds.), 1991. *Biogeochemistry of Major World Rivers*. John Wiley & Sons Ltd.

de Haas, H., T.C.E. van Weering, and H. de Stigter, 2002. Organic carbon in shelf seas: sinks or sources, processes and products. *Continental Shelf Research*, 22:691-717.

Fisher, T.R., D. Correll, R. Constanza, J.T. Hollibaugh, C.S. Hopkinson, R.W., Howarth, N.N. Rabalais, J.E. Richey, C.J. Vorosmarty, and R. Wiegert, 2000. Synthesizing drainage basin inputs to coastal systems. In: *Estuarine Science: A Synthetic Approach to Research and Practice* [Hobbie, J.E. (ed.)]. Island Press, Washington, DC, pp. 81-101.

Fox, L.E., 1991. The transport and composition of humic substances in estuaries. In: *Organic Substances and Sediments in Water* [R.A. Baker (ed.)]. Lewis Publ., Chelsea, MI.

Frankignoulle, M., G. Abril, A. Borges, I. Bourge, C. Canon, B. Delile, E. Libert, and J.-M. Theate, 1998. Carbon dioxide emission from European estuaries. *Science*, 282:434-436.

Frey, K., and L. Smith, 2005. Amplified carbon release from vast West Siberian peatlands by 2100. *Geophysical Research Letters*, 32:L09401.

Geyer, W.R., P.S. Hill, and G.C. Kineke, 2004. The transport, transformation and dispersal of sediment by buoyant coastal flows. *Continental Shelf Research*, 24:927-949.

Goni, M.A., K.C. Ruttenberg, and T.I. Eglinton, 1998. A reassessment of the sources and importance of land-derived organic matter in surface sediments from the Gulf of Mexico. *Geochimica et Cosmochimica Acta*, 62:3055-3075.

Goni, M.A., M.W. Cathey, Y.H. Kim, and G. Voulgaris, 2005a. Fluxes and sources of suspended organic matter in an estuarine turbidity maximum region during low discharge conditions. *Estuarine, Coastal and Shelf Science*, 63:683-700.

Goni, M.A., M.B. Yunker, R.W. Macdonald, and T.I. Eglinton, 2005b. The supply and preservation of ancient and modern components of organic carbon in the Canadian Beaufort Shelf of the Arctic Ocean. *Marine Chemistry*, 93:53-73.

Goni, M.A., N. Monacci, R. Gisewhite, A. Ogston, J. Crockett, and C. Nittrouer, 2006. Distribution and sources of particulate organic matter in the water column and sediments of the Fly River Delta, Gulf of Papua (Papua New Guinea). *Estuarine, Coastal and Shelf Science*, 69:225-245.

Gordon, E.S., and M.A. Goni, 2004. Controls on the distribution and accumulation of terrigenous organic matter in sediments from the Mississippi and Atchafalaya river margin. *Marine Chemistry*, 92:331-352.

Gordon, E.S., M.A. Goni, Q.N. Roberts, G.C. Kineke, and M.A. Allison, 2001. Organic matter distribution and accumulation on the inner Louisiana shelf west of the Atchafalaya River. *Continental Shelf Research*, 21:1691-1721.

Hedges, J.I., J.R. Ertel, P.D. Quay, P.M. Grootes, J.E. Richey, A.H. Devol, G.W. Farwell, F.W. Schmidt, and E. Salati, 1986. Organic carbon-14 in the Amazon River system. *Science*, 231:1129-1131.

Hedges, J., 1992. Global biogeochemical cycles: Progress and problems. *Marine Chemistry*, 39:67-93.

Hedges, J.I., and R.G. Keil, 1995. Sedimentary organic matter preservation: an assessment and speculative synthesis. *Marine Chemistry*, 49:81-115.

Hedges, J.I., R.G. Keil, and R. Benner, 1997. What happens to terrestrial organic matter in the ocean? *Organic Geochemistry*, 27:195-212.

Hickey, B.M., and N.S. Banas, 2003. Oceanography of the US Pacific Northwest Coastal Ocean and estuaries with application to coastal ecology. *Estuaries*, 26:1010-1031.

Hobbie, J.E. (ed.), 2000. *Estuarine Science: A Synthetic Approach to Research and Practice*. Island Press, Washington, DC, 539 pp.

Hoffman, J.C., and D.A. Bronk, 2006. Interannual variation in stable carbon and nitrogen isotope biogeochemistry of the Mattaponi River, Virginia. *Limnology and Oceanography*, 51:2319-2332.

Hopkinson, C.S., et al. 1998. Terrestrial inputs of organic matter to coastal ecosystems: an intercomparison of chemical characteristics and bioavailability. *Biogeochemistry*, 43:211-234.

Howarth, R.W., G. Billen, D. Swaney, A. Townsend, N. Jaworski, K. Lajtha, J. A. Downing, R. Elmgren, N. Caraco, T. Jordan, F. Berendse, J. Freney, V. Kudeyarov, P. Murdoch and Z. Zhao-Liang, 1996. Regional nitrogen budgets and riverine N & P fluxes for the drainages to the North Atlantic Ocean: Natural and human influences. *Biogeochemistry*, 35:75-139.

Humborg, C., 1997. Primary productivity regime and nutrient removal in the Danube Estuary. *Estuarine, Coastal and Shelf Science*, 45:579-589.

Humborg, C., D.J. Conley, L. Rahm, F. Wulff, A. Cociasu, and V. Ittekkot, 2000. Silica retention in river basins: far-reaching effects on biogeochemistry and aquatic food webs in coastal marine environments. *Ambio*, 29:45-50

Ittekkot, V., 1988. Global trends in the nature of organic matter in river suspensions. *Nature*, 332:436-438.

Ittekkot, V., and R.W.P.M. Laane, 1991. Fate of riverine particulate organic matter. In: *Biogeochemistry of Major World Rivers* [Degens, E.T., S. Kempe, and J.E. Richey (eds.)]. John Wiley & Sons Ltd.

Ittekkot, V., C. Humborg, and P. Schafer, 2000. Hydrological alterations and marine biogeochemistry: A silicate issue? *Bioscience*, 50:776-782.

Jay, D.A., W.R. Geyer, and D.R. Montgomery, 2000. An ecological perspective on estuarine classification. In: *Estuarine Science: A Synthetic Approach to Research and Practice* [J.E. Hobbie (ed.)]. Island Press, Washington, DC, pp. 149-176.

Jickells, T.D., 1998. Nutrient biogeochemistry of the coastal zone. *Science*, 281:217-222.

Kieber, R.J., X. Zhou, and K. Mopper, 1990. Formation of carbonyl compounds from UV-induced photodegradation of humic substances in natural waters: Fate of riverine carbon in the sea. *Limnology and Oceanography*, 35:1503-1515.

Kineke, G.C., R.W. Sternberg, J.H. Trowbridge, and W.R. Geyer, 1996. Fluid mud processes on the Amazon continental shelf. *Continental Shelf Research*, 16:667-696.

Kineke, G.C., K.J. Woolfe, S.A. Kuehl, J.D. Milliman, T.M. Dellapenna, and R.G. Purdon, 2000. Sediment export from the Sepik River, Papua New Guinea: evidence for a divergent sediment plume. *Continental Shelf Research*, 20:2239-2266.

Leithold, E.L., and R.S. Hope, 1999. Deposition and modification of a flood layer on the northern California shelf: lessons from and about the fate of terrestrial particulate organic carbon. *Marine Geology*, 154:183-195.

Leithold, E.L., D.W. Perkey, N.E. Blair, and T.N. Creamer, 2005. Sedimentation and carbon burial on the northern California continental shelf: the signatures of land-use change. *Continental Shelf Research*, 25:349-371.

Lohrenz, S.E., and W.-J. Cai, 2006. Satellite ocean color assessment of air-sea fluxes of CO_2 in a river-dominated coastal margin. *Geophysical Research Letters*, 33:L01601, doi:10.1029/2005GL023942.

Ludwig, W., and J.L. Probst, 1998. River sediment discharge to the global oceans: Present-day controls and global budgets. *American Journal of Science*, 298:265-295.

Ludwig, W., J.-L. Probst, and S. Kempe, 1996. Predicting the oceanic input of organic carbon by continental erosion. *Global Biogeochemical Cycles*, 10:23-41.

Masiello, C.A., and E.R.M. Druffel, 2001. Carbon isotope geochemistry of the Santa Clara River. *Global Biogeochemical Cycles*, 15:407-416.

Mayorga, E., A.K. Aufdenkampe, C.A. Masiello, A.V. Krusche, J.I. Hedges, P.D. Quay, J.E. Richey, and T.A. Brown, 2005. Young organic matter as a source of carbon dioxide outgassing from Amazonian rivers. *Nature*, 436:538-541.

McKee, B., 2003. RiOMar: the transport, transformation and fate of carbon in river-dominated ocean margins. *Report of the RiOMar Workshop*, 1-3 November 2001, Tulane University, New Orleans.

McKee, B., 2005. River-dominated Ocean Margins (RiOMar): Linkages with global climate change. *OCB Workshop Proceedings*, Woods Hole, MA, July 2005.

McKee, B.A., R.C. Aller, M.A. Allison, T.S. Bianchi, and G.C. Kineke, 2004. Transport and transformation of dissolved and particulate materials on continental margins by major rivers: benthic boundary layer and seabed processes. *Continental Shelf Research*, 24:899-926.

Meade, R.H., T.R. Yuzyk, and T.J. Day, 1990. Movement and storage of sediment in rivers of the United States and Canada. In: *The Geology of North America*, v. O-1 [M. Wolman and H. Riggs (eds.)]. Geological Society of America, Boulder, CO, pp. 255-280.

Meybeck, M., 1982. Carbon, nitrogen, and phosphorus transport by world rivers. *American Journal of Science*, 282:401-450.

Meybeck, M., 1993. Riverine transport of atmospheric carbon: sources, global typology, and budget. *Water, Air and Soil Pollution*, 70:443-463.

Meybeck, M., and C. Vorosmarty, 1999. Global transfer of carbon by rivers. *IGBP Global Change Newsletter*, 37:18-19.

Meybeck, M., and C. Vorosmarty, 2005. Fluvial filtering of land-to-ocean fluxes: From natural Holocene variations to Anthropocene. *Comptes Rendus Geoscience*, 337:107-123.

Minor, E.C., J.J. Boon, H.R. Harvey, and A. Mannino, 2001. Estuarine organic matter composition as probed by direct temperature-resolved mass spectrometry and traditional geochemical techniques. *Geochimica et Cosmochimica Acta*, 65:2819-2834.

Mopper, K., and D.J. Kieber, 2002. Photochemistry and the cycling of carbon, sulfur, nitrogen and phosphorus. In: *Biogeochemistry of Marine Dissolved Organic Matter* [D. Hansell and C. Carlson (eds.)]. Academic Press, San Diego.

Moran, M.A., and R.E. Hodson, 1994. Support of bacterioplankton production by dissolved humic substances from three marine environments. *Marine Ecology Progress Series*, 110:241-247.

Moran, M.A., R.J. Wicks, and R.E. Hodson, 1991. Export of dissolved organic matter from a mangrove swamp ecosystem: Evidence from natural fluorescence, dissolved lignin phenols, and bacterial secondary production. *Marine Ecology Progress Series*, 76:175-184.

Moran, M.A., W.M. Sheldon, and R.G. Zepp, 2000. Carbon loss and optical property changes during long-term photochemical and biological degradation of estuarine dissolved organic matter. *Limnology and Oceanography*, 45:1254-1264.

Mulholland, P.J., and J.A. Watts, 1982. Transport of organic carbon to the oceans by rivers of North America: A synthesis of existing data. *Tellus*, 34:176-186.

Neff, J.C., S.A. Finlay, S.P. Zimov, J.J. Davydov, E.A. Carrasco, G. Schuur, and A.I. Davydova, 2006. Seasonal changes in the age and structure of dissolved organic carbon in Siberian rivers and streams, *Geophysical Research Letters*, 33:L23401, doi:10.1029/2006GL028222.

Nittrouer, C.A., and L.D. Wright, 1994. Transport of particles across continental shelves. *Reviews of Geophysics*, 32:85-113.

Nittrouer, C.A., S.A. Kuehl, A.G. Figueiredo, M.A. Allison, C.K. Sommerfield, J.M. Rine, L.E.C. Faria, and O.M. Silveira, 1996. The geological record preserved by Amazon shelf sedimentation. *Continental Shelf Research*, 16:817-841.

Nittrouer, C.A., S. Miserocchi, and F. Trincardi, 2004. The PASTA project: Investigation of Po and Apennine sediment transport and accumulation. *Oceanography*, 17:46-57.

Ogston, A.S., D.A. Cacchione, R.W. Sternberg, and G.C. Kineke, 2000. Observations of storm and river flood-driven sediment transport on the northern California continental shelf. *Continental Shelf Research*, 20:2141-2162.

Opsahl, S., and R. Benner, 1997. Distribution and cycling of terrigenous dissolved organic matter in the ocean. *Nature*, 386:480-482.

Opsahl, S., and R.G. Zepp, 2001. Photochemically-induced alteration of stable carbon isotope ratios ($\delta^{13}C$) in terrigenous dissolved organic carbon. *Geophysical Research Letters*, 28:2417-2420.

Peterson, B.J., R.M. Holmes, J.W. McClelland, C.J. Vörösmarty, R.B. Lammers, A.I. Shiklomanov, I.A. Shiklomanov, and S. Rahmstorf, 2002. Increasing river discharge to the Arctic Ocean. *Science*, 298:2171-2173.

Pocklington, R., and F.C. Tan, 1987. Seasonal and annual variations in the organic matter contributed by the St. Lawrence River to the Gulf of St. Lawrence. *Geochimica et Cosmochimica Acta*, 51:2579-2581

Postel , S., G. Daily, and P. Ehrlich, 1996. Human appropriation of renewable fresh water. *Science*, 271:785-788.

Prahl, F.G., J.M. Hayes, and T.-M. Xie, 1992. Diploptene: An indicator of terrigenous organic carbon in Washington coastal sediments. *Limnology and Oceanography*, 37:1290-1300.

Prahl, F.G., J.R. Ertel, M.A. Goni, M.A. Sparrow, and L.A. Pinto, 1994. Terrestrial organic carbon contributions to sediments on the Washington margin. *Geochimica et Cosmochimica Acta*, 58:14.

Raymond, P.A., and J.E. Bauer, 2000. Bacterial consumption of DOC during transport through a temperate estuary. *Aquatic Microbial Ecology*, 22:1-12.

Raymond, P.A., and J.E. Bauer, 2001a. Riverine export of aged terrestrial organic matter to the North Atlantic Ocean. *Nature*, 409:497-500.

Raymond, P.A., and J.E. Bauer, 2001b. Coupled ^{14}C and ^{13}C natural abundances as a tool for evaluating freshwater, estuarine and coastal organic matter sources and cycling. *Organic Geochemistry*, 32:469-485.

Raymond, P.A., and J.E. Bauer, 2001c. DOC cycling in a temperate estuary: a mass balance approach using natural ^{14}C and ^{13}C. *Limnology and Oceanography*, 46:655-667.

Raymond, P.A., and J.J. Cole, 2003. Increase in the export of alkalinity from North America's largest river. *Science*, 301:88-91.

Raymond, P.A., and C.S. Hopkinson, 2003. Ecosystem modulation of dissolved carbon age in a temperate marsh-dominated estuary. *Ecosystems*, 6:694-705.

Raymond, P.A., J.E. Bauer, N. Caraco, J.J. Cole, B. Longworth, and S. Petsch, 2004. Controls on the variability of organic and dissolved inorganic carbon age in northeast U.S. rivers. *Marine Chemistry*, 92:353-366.

Richey, J.E., J.I. Hedges, A.H. Devol, P.D. Quay, R. Victoria, L. Martinelli, and B.R. Forsberg, 1990. Biogeochemistry of carbon in the Amazon River. *Limnology and Oceanography*, 35:352-371.

Richey, J.E., J.M. Melack, A.K. Aufdenkampe, V.M. Ballester, and L.L. Hess, 2002. Outgassing from Amazonian rivers and wetlands as a large tropical source of atmospheric CO_2. *Nature*, 416:617-620.

Schlunz, B., and R. R. Schneider, 2000. Transport of terrestrial organic carbon to the oceans by rivers: Re-estimating flux and burial rates. *International Journal of Earth Sciences*, 88:599-606.

Serageldin, I., 1995. *Toward Sustainable Management of Water Resources*. The World Bank, Washington, DC.

Socolow, R.H., 1999. Nitrogen management and the future of food: lessons from the management of energy and carbon. *Proceedings of the National Academy of Sciences*, 96:6001-6008.

Spitzy, A., and V. Ittekkot, 1991. Dissolved and particulate organic matter in rivers. In: *Ocean Margin Processes in Global Change*. Report, Dahlem Workshop, Berlin, [R.F.C. Mantoura (ed.)], pp. 5-17.

Spitzy, A., and J. Leenheer, 1991. Dissolved organic carbon in rivers. In: *Biogeochemistry of Major World Rivers* [E.T. Degens, S. Kempe, and J.E. Richey (eds.)]. John Wiley & Sons Ltd., pp. 213-232.

Stein, R., and R.W. Macdonald (eds.), 2004. *The Organic Carbon Cycle in the Arctic Ocean*. Springer-Verlag, 363 pp.

Striegl, R.G., G.R. Aiken, M.M Dornblaser, P.A. Raymond, and K.P. Wickland, 2005. A decrease in the discharge normalized DOC export by the Yukon River during summer through autumn. *Geophysical Research Letters*, 32:L21413, doi:10.1029/2005GL024413.

Sullivan, B.E., F.G. Prahl, L.F. Small, and P.A. Covert, 2001. Seasonality of phytoplankton production in the Columbia River: A natural or anthropogenic pattern? *Geochimica et Cosmochimica Acta*, 65:1125-1139.

Ternon, J.F., C. Oudot, A. Dessier, and D. Diverres, 2000. A seasonal tropical sink for atmospheric CO_2 in the Atlantic ocean: the role of the Amazon River discharge. *Marine Chemistry*, 68:183-201.

Thorp, J.H., and M.D. Delong, 1994. The river production model: An heuristic view of carbon sources and organic processing in large river ecosystems. *Oikos*, 70:305-308.

Ver, L.M.B., F.T. Mackenzie, and A. Lerman, 1999. Carbon cycle in the coastal zone: Effects of global perturbations and change in the past three centuries. *Chemical Geology*, 159:283-304.

Vorosmarty, C.J., and B.J. Peterson, 2000. Macro-scale models of water and nutrient flux to the coastal ocean. In: *Estuarine Science: A Synthetic Approach to Research and Practice* [J.E. Hobbie (ed.)]. Island Press, Washington, DC, pp. 43-79.

Wetzel, R.G., 2001. *Limnology: Lake and River Ecosystems*. Academic Press.

Wheatcroft, R.A., and J.C. Borgeld, 2000. Oceanic flood sedimentation: a new perspective. *Continental Shelf Research*, 20:2059-2066.

Wysocki, L.A., T.S. Bianchi, R.T. Powell, and N. Reuss, 2006. Spatial variability in the coupling of organic carbon, nutrients, and phytoplankton pigments in surface waters and sediments of the Mississippi River plume. *Estuarine, Coastal and Shelf Science*, 69:47-63.

Observation and Synthesis of Carbon Cycling on the Continental Margins

Oscar Schofield
Institute of Marine and Coastal Science
Rutgers University

Lou Codispoti and Vince Kelly
Horn Point Laboratory
University of Maryland

Mike DeGrandpre
University of Montana

Niki Gruber
Institute of Biogeochemistry and Pollutant Dynamics
Department of Environmental Sciences
ETH Zürich

Dave Siegel
University of California, Santa Barbara

Wade McGillis
Lamont Doherty Earth Observatory
Columbia University

Chris Sabine
Pacific Marine Environmental Laboratory
National Oceanic and Atmospheric Administration

The previous sections demonstrated extreme dynamic ranges of variability in ocean-margin carbon-relevant parameters over a wide range of length- and time scales. Without automated, high-resolution sampling and analysis/sensing procedures, it is difficult to capture the variability inherent in coastal systems within reasonable time and expense constraints. While many oceanographic measurement programs have valued high accuracy, precision, and sensitivity over measurement-density, high analysis speed, and low cost, adequately sampling the coastal ocean may require a re-thinking of that paradigm. Along with high variability in the parameters we desire to measure comes extreme variability in 'conditional' parameters (e.g., T, S, turbidity). This increases concerns over measurement interferences, as changes in a target parameter will be coincident with changes in other factors that may influence measurement response. The focus may shift from increased sampling precision and stability to more rapid-response analyses, and smaller, less expensive instrumentation that could be deployed on a wider variety of platforms by a wider user group. Additional emphasis may be placed on remote-sensing products, and their relationships with other relevant parameters, to expand spatial and temporal coverage of surface waters. Further, models are expected to play a major role in coastal carbon-cycle studies, both from the perspective of understanding conditions in under-sampled regions and that of predicting future responses of these systems to changing climates.

In-water measurement approaches—Ship-based observations

Sensing and analyses of waters drawn through ships' surface intake lines have become more expansive, now regularly including sensor-based measurements of dissolved O_2 and a host of bio-optical measurements (fluorescence, beam attenuation, CDOM fluorescence) that all may have relevance to surface CO_2 distributions. Continuous surface-underway measurements of pCO_2 are now routine enough that equilibrator/infra-red gas analyzer systems have been deployed aboard container ships (e.g., Takahashi et al. 2002) and cruise liners (Wanninkhof et al. 2007) without attendant technical personnel, and a few of the larger research vessels now routinely incorporate these systems in their surface underway sample lines. With a few exceptions (e.g., Hales et al. 2004b), these CO_2-measurement systems are large, have response times of several minutes, and are quite expensive to construct. These features make them less than ideal for widespread deployment on smaller coastal vessels where signals change rapidly.

Other analyses and sensing techniques have now been automated and adapted for continuous measurement of relevant parameters in flowing streams of seawater, including nutrients (Johnson and Coletti, 2002; Wetz et al. 2006; Hales et al. 2004a), pCO_2 (using either the showerhead- (Broecker and Takahashi, 1966; Weiss, 1974) or hydrophobic membrane- (Hales et al. 2004b) style equilibrators), total CO_2 (Kaltin et al. 2005; Bandstra et al. 2006; Hiscock et al. 2006), alkalinity (Roche and Millero, 1998), and pH (e.g., Byrne and Breland, 1989). Only a few such systems (Johnson and Coletti, 2002; Wetz et al. 2006; Hales et al. 2004a, b; Bandstra et al. 2006) have the fast response times desired for work in highly variable coastal systems. None of these (with the possible exception of the commercially available ISUS units) are, as yet, robust enough for operation without skilled personnel on board the ship. All are still quite expensive to purchase or fabricate.

With few exceptions, ship-based study of the coastal oceans below the sea surface is still largely centered on traditional approaches to sampling small numbers of discrete depths at discrete locations, followed by discrete analyses of samples. Among the high-resolution vertically resolved surveys in North American coastal waters, only a few (Figure 9.1; Hales et al. 2005a, b, 2006; Bandstra et al. 2006) have included detailed chemical measurements. These approaches, while powerful, are technically challenging.

In-water measurement approaches—mooring-based observations

Coastal moorings present an underutilized deployment opportunity for coastal observations. There are thousands of navigational and weather buoys in North American coastal waters, some of which

Figure 9.1. Example of ship-based high-resolution vertical sections of pCO_2 across the shelf off Oregon. (From Hales et al. 2005a).

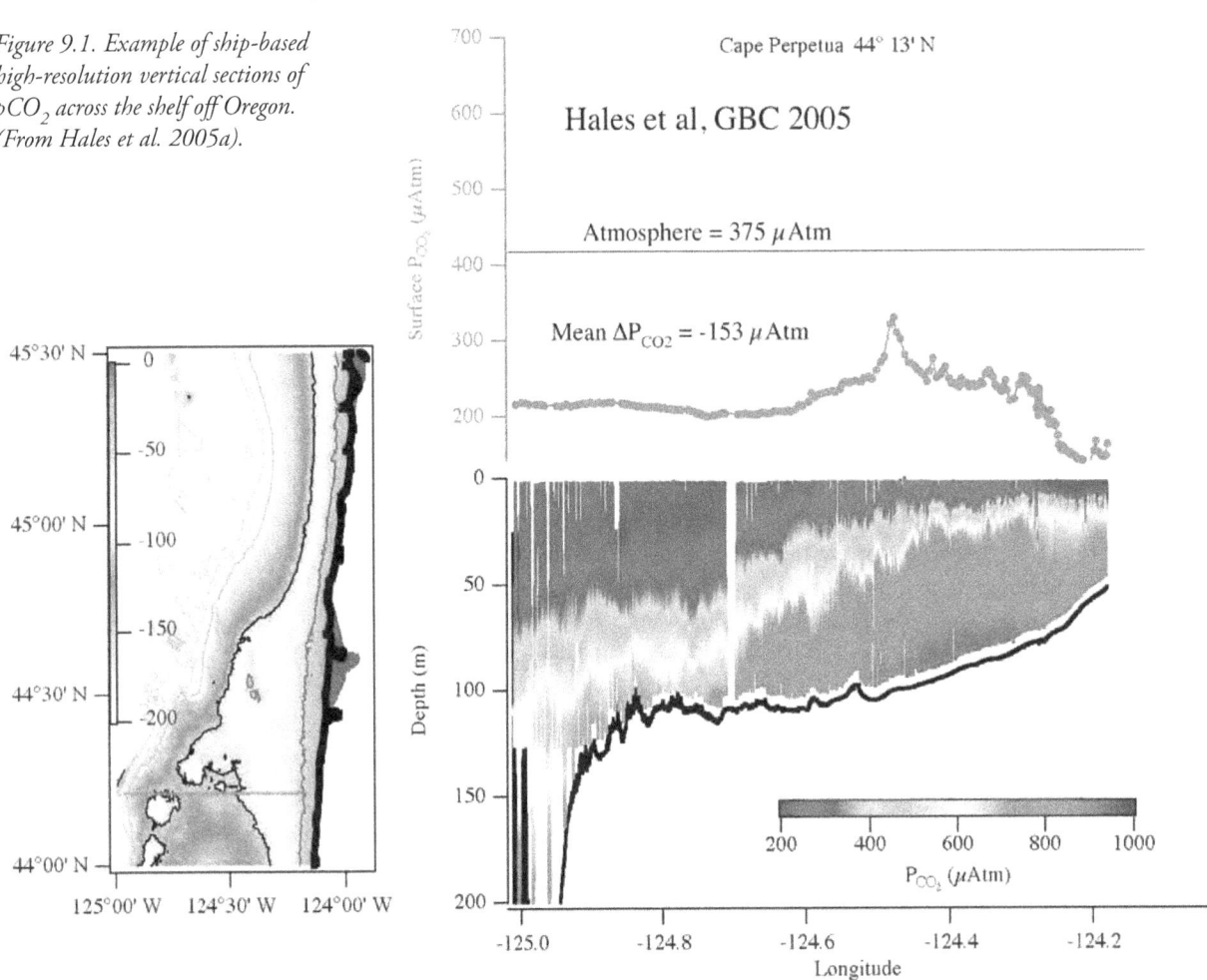

already include basic hydrographic measurements (e.g., SST on NDBC buoys). These platforms were not designed specifically to be heavily outfitted with oceanographic instrumentation, however, and there are only a few instances of chemical sensors being deployed on weather (C. Sabine, pers. comm.; http://www.pmel.noaa.gov/co2/coastal) or navigation buoys (e.g., Sigleo et al. 2005). As a result, measurement systems that could be deployed on these platforms probably need to be smaller than would be the case for larger oceanographic mooring platforms. Further, given the large number of moorings of opportunity and the high spatial variability of the ocean margins, the costs of individual systems should not prohibit broad deployment. Options for mooring-based measurements are slightly more limited than ship-based options. A multitude of bio-optical sensors have been deployed in a variety of settings: there are now sensors for O_2 ((http://www.seabird.com; www.aanderaa.com) and nitrate (http://www.satlantic.com) with sufficient stability that they can be programmed to recover data autonomously on a mooring for several-month durations, and there are a couple of options for pCO_2 measurement. There are fewer viable options for other chemical measurement systems.

One of these pCO_2 systems, developed by DeGrandpre et al. (1995) (Figure 9.2), is based on detection of color changes of a pH-sensitive dye caused by diffusion of CO_2 across a gas-permeable membrane, and is commercially available (http://www.sunburstsensors.com). This system has the advantage that it can be deployed at depths of up to a few hundred meters, thus providing some information about vertical pCO_2 distributions. The other, developed jointly by researchers at PMEL and MBARI (Figure 9.2), is based on IR detection of the CO_2 content of a gaseous sample equilibrated with surface waters. This system has the advantages that it uses measurement technology common with that typically used for shipboard analyses, can be directly calibrated with standard gas mixtures, and can incorporate regular determinations of the CO_2 content of marine boundary-layer

air as part of its normal measurement schedule. It has the disadvantages of being deployable only on the surface expression of a mooring, and not being directly commercially available.

Moored analyzers for nutrients in addition to nitrate are commercially available (http://www.oceanmarineinc.com; http://www.ysi.com) and have been successfully deployed in a few locations (Sigleo et al. 2005; Glibert et al. 2005; Figure 9.3), but these have proven to require dedicated, skilled technical personnel to successfully operate in the field. Other autonomous

Figure 9.2. Moored observations of pCO$_2$ off the a) US Atlantic coast, using a colorimetric system (from DeGrandpre et al. 2002), and b) Pacific coast, using a NDIR-based system (C. Sabine, pers. comm.; http://www.pmel.noaa.gov/co2/coastal).

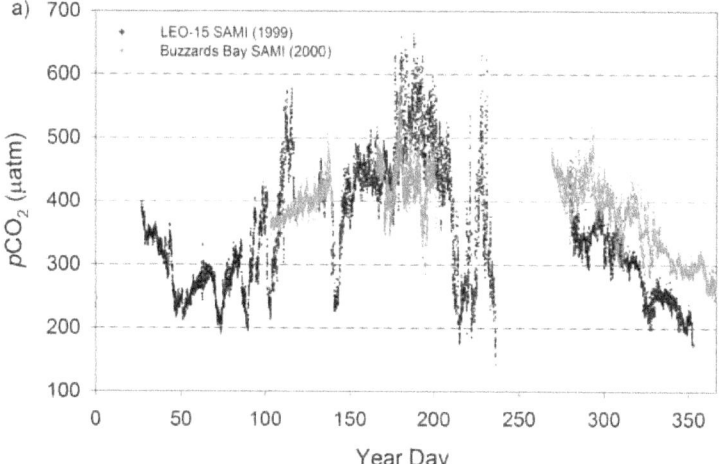

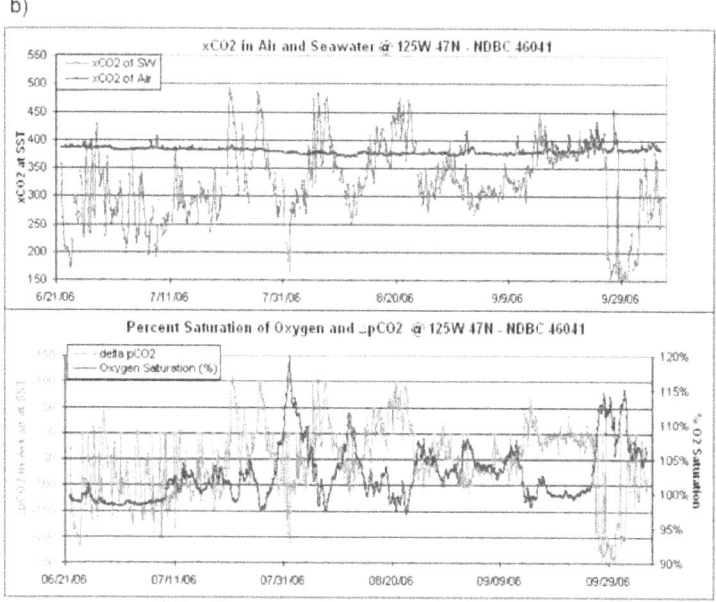

http://www.pmel.noaa.gov/co2/coastal

analyzers for pH, alkalinity, and trace metals are in the developmental stage (M. Degrandpre, pers. comm.; Holm, 2006).

Aside from the more basic bio-optical sensors, all of these instruments are costly and bulky (Table 9.1). Outfitting a single mooring with systems for measurement at five depths of, e.g., pCO_2, nitrate, and silicate, in addition to the more routine measurements of temperature, salinity, fluorescence, O_2, and optical beam attenuation, would cost several hundred thousand dollars just for the sensors themselves (Table 9.1). This presents a significant barrier to populating the coastal ocean with instrumented moorings at any spatial density that could hope to capture the inherent variability of the system. The size of these instruments requires large mooring platforms, typically deployed only from the larger vessels in the UNOLS fleet. This makes it difficult to adequately sample shallow water areas where these vessels do not routinely operate, and adds a level of cost and complexity to mooring deployment that further limits attainment of the objective of densely-spaced moorings.

In-water measurements—submersible vehicle-based observations

Autonomous vehicle platforms are just making the transition to an operational status, and present the combined requirements of fast response needed for ship-based systems, and low-power requirements of moored systems, with further restrictions on payloads. The United States Navy is beginning to actively incorporate these platforms into their environmental mapping capabilities. Such platforms include unmanned tethered ROVs (http://www.whoi.edu/marops/vehicles/index.html), AUVs (http://www.bluefinrobotics.com; Blackwell et al. 2002; Moline et al. 2005) and free-vehicle gliders (Eriksen et al. 2001; Rudnick et al. 2004; Glenn et al. 2004; Schofield et al. 2007). Size, power requirements, and response times of available sensors limit their utility on these platforms, although measurements of temperature, salinity, currents, ocean turbidity, and oxygen and chlorophyll concentrations are currently routine (Figure 9.4). These systems can also be flown as fleets to provide high-

Figure 9.3. Left: Time-series of nitrate data collected using an autonomous nitrate analyzer, deployed from a navigational buoy off the central Oregon Coast. From Sigleo et al. (2005). Right: Time series of nitrate plus nitrite (upper panel), reactive phosphorus (middle panel), and salinity and daily total rainfall (lower panel) in the mouth of the Pocomoke River, a tributary of Chesapeake Bay. Nutrient data was collected with W.S. Ocean systems in-situ monitors, salinity was measured with a Sea-Bird MicroCAT, and rainfall data was collected at a USGS station in Snow Hill, MD. All instruments were mounted on a Chesapeake Bay Observing System Buoy (CBOS) at an approximate depth of 1m. (From Glibert et al. 2005).

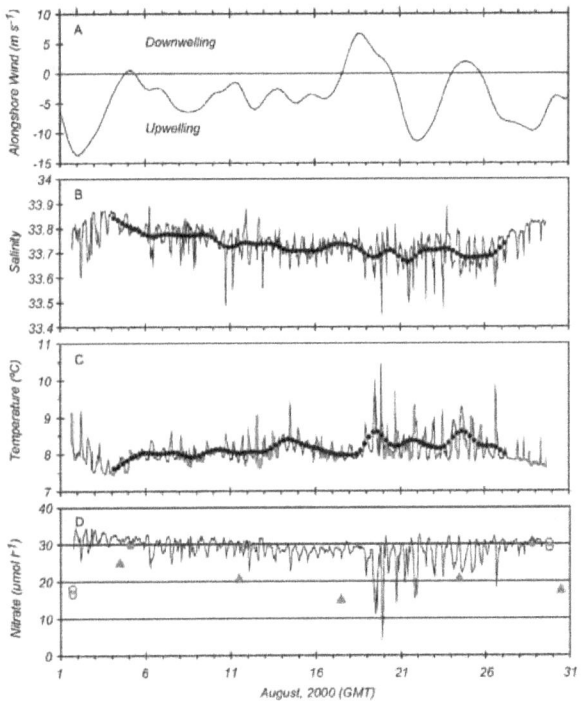

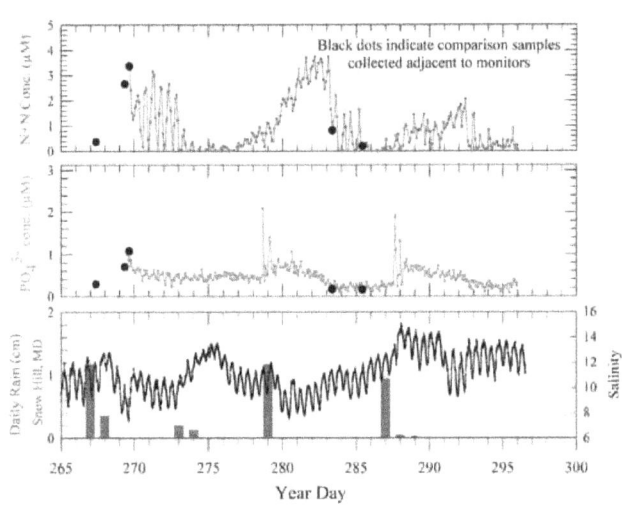

Table 9.1. Summary of cost and availability of mooring-deployable instrumentation.

Measured parameter	Measurement approach	Manufacturer	Approximate cost
Temperature, Salinity, Depth, datalogger	Electronic sensors	Various	$15k
$p\mathrm{CO_2}$, subsurface	Optical, based on color change of pH-sensitive solution	Sunburst Sensors, LLC	$20k
$p\mathrm{CO_2}$ surface only	NDIR of equilibrated gaseous headspace	N/A; manufactured at MBARI and PMEL	$20k
Nitrate	Direct UV absorbance	Satlantic	$25k
Nitrate, phosphate, silicate, nitrate, ammonium	Wet-chemistry followed by optical detection	Various	~$12k per analysis
Chlorophyll, POC, CDOM	Passive or active optical detection	Various	~4k per parameter
Dissolved $\mathrm{O_2}$	Electrochemical or optical (Fluorescence quenching)	SeaBird, Anderaa	$5k
Complete suite of measurements at a single depth, including CTD, $p\mathrm{CO_2}$, nitrate, phosphate, silicate, $\mathrm{O_2}$, chlorophyll, CDOM, and POC			~$100k

Figure 9.4. Data collected at Martha's Vineyard Cabled Observatory during a tropical storm in fall 2005. The data were collected during a single transect highlighted in the map. The data provided by this glider included temperature, salinity, chlorophyll fluorescence, colored dissolved organic matter (CDOM) fluorescence, and light backscatter at 23 wavelengths. The lack of correlation between chlorophyll-a fluorescence and light backscatter reflects the sediment resuspension in the bottom waters during the storm. Data collected and provided from the Office of Naval Research OASIS program by Michael Twardowski.

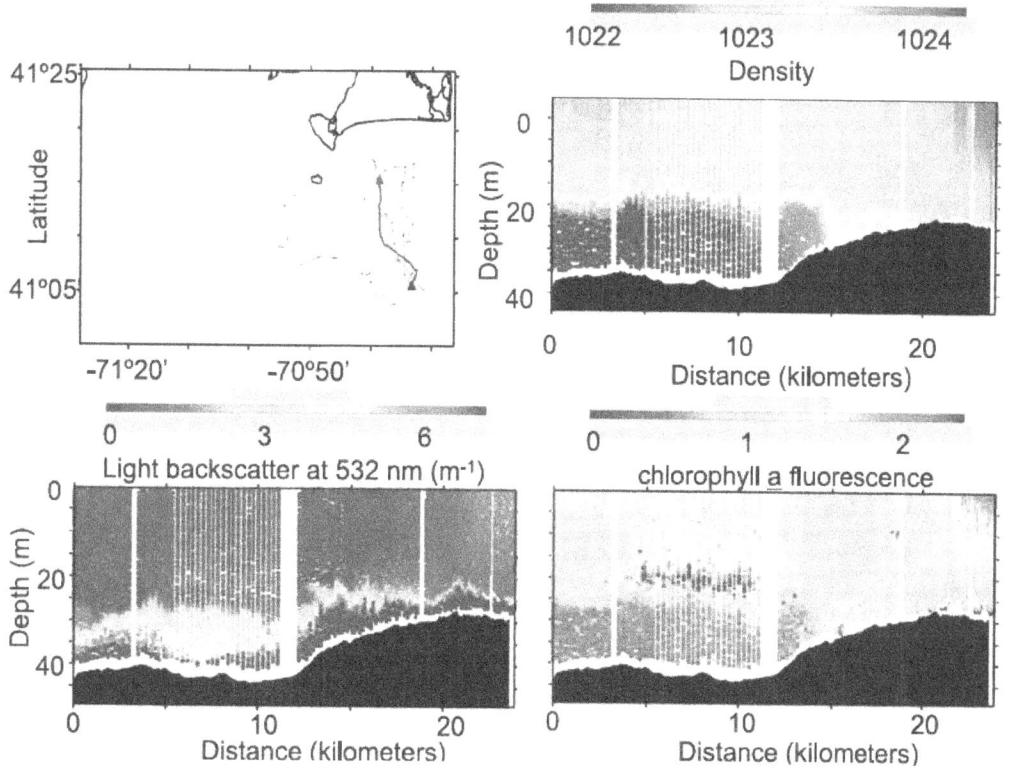

resolution volumes of data in near real-time to enable ship-based process studies (Figure 9.5). The advantage of the systems is their ability to collect spatial data even during extreme events that may play disproportionately large roles in coastal carbon cycles. Miniaturization of the submersible nutrient sensors is underway. The development of small pCO_2 sensors should be pursued as the systems have demonstrated the capability to maintain a sustained presence at sea under extreme conditions.

In-water measurements—Seafloor observations

Requirements for measurements of seafloor properties and processes contain elements of those for ship-based, mooring-based, and free-vehicle-based observations, as benthic research often takes place via the short-term (hours-weeks) deployment from research vessels of autonomous instrumentation to the seafloor. The study of benthic processes is likewise limited, with most coastal benthic-process researchers deploying instrumentation and techniques originally designed for the more-benign deep-ocean environment. A few have developed systems for better two-dimensional and temporal coverage of benthic environments (Wheatcroft, 1994; Berg et al. 2003; Glud et al. 1999). One distinct advantage of working on the seafloor in the ocean margins where sediment diagenesis is rapid and redox horizons are compressed in the sediment water column is the availability of natural power derived from these gradients (e.g., Reimers et al. 2001, 2006). Again, these tools are far from simple and are nowhere near the plug-and-play status that would be required for their widespread deployment by coastal researchers who may lack technical expertise with these systems.

Figure 9.5. An example of salinity data collected by a fleet of six gliders at the shelf-slope interface on the Mid-Atlantic Bight. The data was collected as part of the Office of Naval Research Shallow Water 2006 and Non-Linear Wave Interactions programs. Figure provided by Dong Lai Gong and Scott Glenn (Rutgers University).

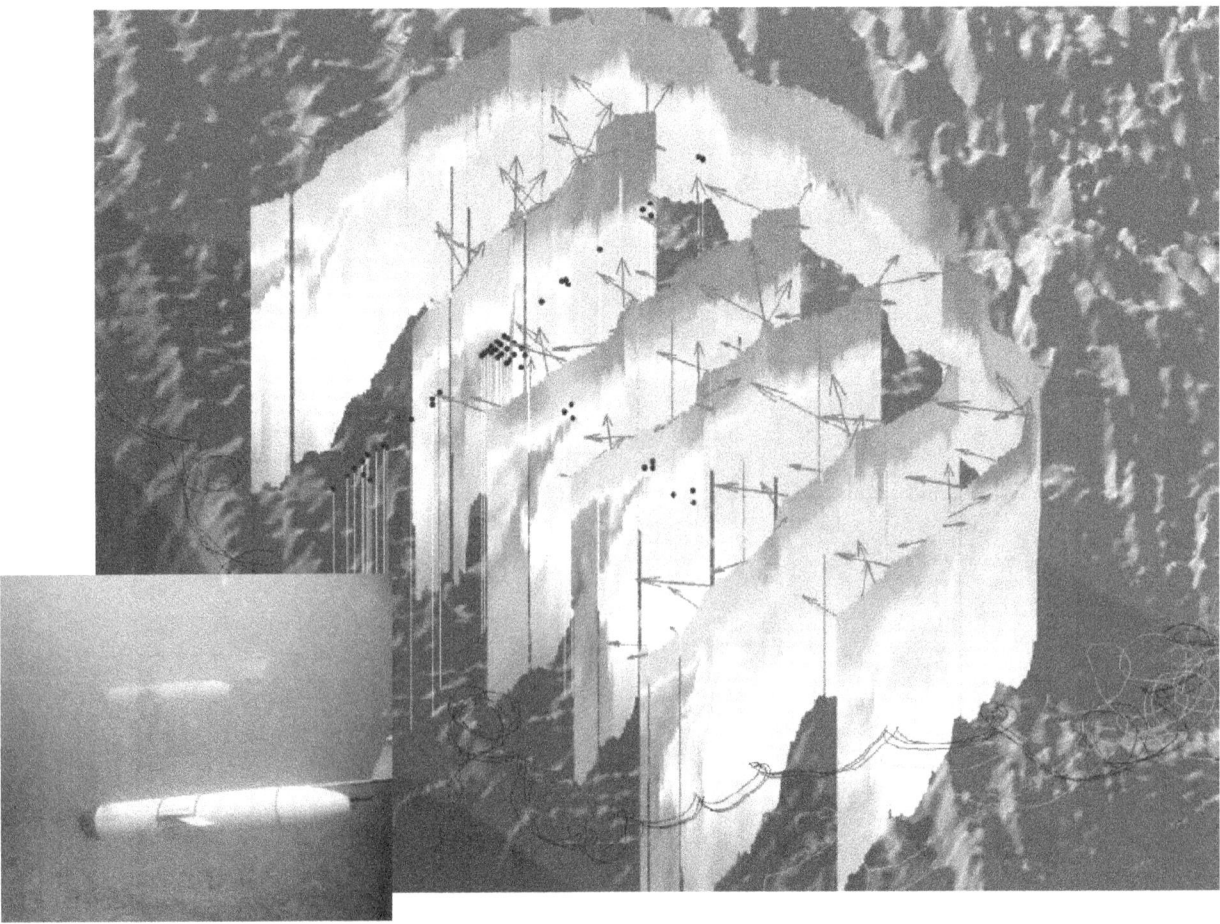

In-water measurements—Ocean Observatories

The ocean observatories are still maturing (Figure 9.6) but will provide a potential backbone in which to integrate sensors that are available for ship, moored, and autonomous-vehicle deployments from the seafloor to the sea surface. The more mature ocean observatories now routinely include physical and bio-optical measurements for extended periods of time. Nutrient sensors, imaging flow cytometers, and occasionally pCO_2 are just beginning to be deployed. Like the mooring and ships, the more advanced chemical and biological sensors require expert personnel; however given current plans to develop a national network of ocean observing systems, any coastal carbon program should leverage off these assets. Several large observing network ventures are under varying states of development within the coastal waters of the United States. These networks will provide scientists an interactive capability to communicate with sensors moored in the sea. The Ocean Research Interactive Observatory Networks (ORION) is a National Science Foundation effort (http://www.orionprogram.org/PDFs/OOI_Science_Plan.pdf, http://www.orionprogram.org/PDFs/workshop_report.pdf) to be initiated within the next five years. A complementary system is the Integrated Ocean Observing System (IOOS) being designed to provide a national backbone of ocean observations for the United States (http://www.ocean.us/documents/docs/Core_lores.pdf). These systems will be a component of the larger international Global Ocean Observing System (GOOS). The advent of these many initiatives offers great potential for the scientific community to conduct holistic system integrated experiments.

Remote Sensing-based observations

The availability of satellite-based measurements of ocean temperature, sea-surface height, surface winds, and color revolutionized our ability to map ocean conditions, circulation patterns, and productivity. While limited to ocean surface waters, remote-sensing measurements present otherwise unattainable two-dimensional coverage at near-daily frequencies. Like other observational techniques, however, remote-sensing approaches are most highly-developed for the open oceans. Algorithms were developed primarily

for "Case 1" waters where the optical properties were dominated by water molecules and phytoplankton allowing relatively robust algorithms for phytoplankton biomass and productivity to be developed. Remote sensing on the continental shelves, however, is more difficult as the waters on shelves tend to be optically complex, representing the variable contributions of phytoplankton, colored dissolved organic matter,

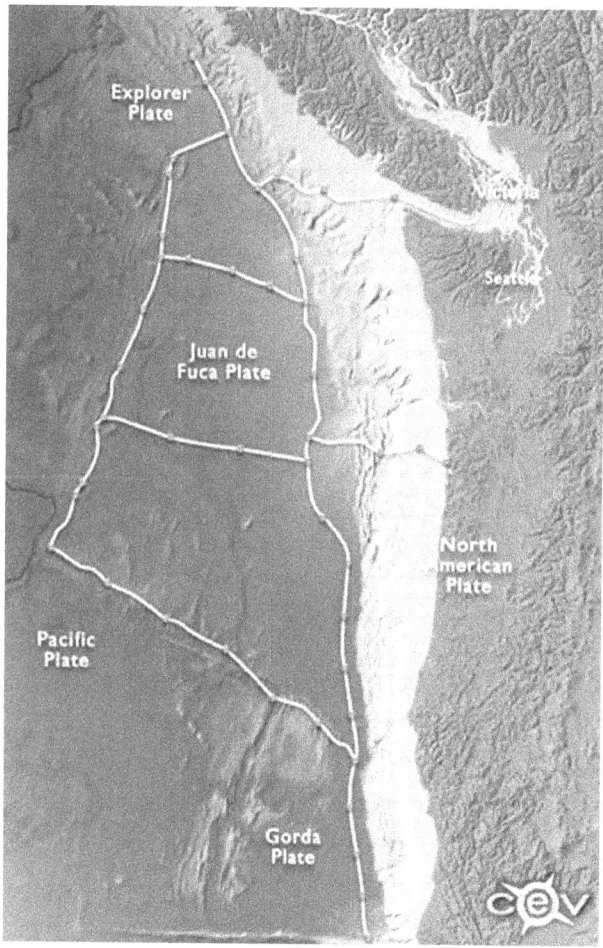

Figure 9.6. The Neptune system is being jointly developed and deployed by Canada and the United States. The fiber optic cable will be outfitted with a series of nodes that will allow instruments to be plugged in for power and shore-based control of the instrument. There will be real-time data telemetry. The Neptune Canada portion of the system is being installed, leading the way to the US contribution through the ORION program. Many different observatory systems are currently being developed world-wide and in the near future will allow scientists to maintain a continuous presence at sea 24 hours a day, 365 days a year. The figure was provided courtesy of the Neptune team at the University of Washington.

sediment, and detritus. Additionally, many of the critical features on shelves are highly dynamic, changing on the time scale of hours, and can be smaller than the approximately kilometer-scale resolution of most satellite systems. Further, complex atmospheric conditions at the ocean-continent boundaries complicate extricating quantitative reflectance values.

Despite these hurdles, significant progress has been achieved over the last decade, including more robust algorithms that are appropriate for coastal waters, satellites with improved spatial resolution (<1 km; Figure 9.7), and improved temporal resolution if the full international constellation of ocean color satellites is accessed. Given these advances, remote sensing will be a critical component of any coastal carbon program. Improvements in several critical areas, however, would greatly enable coastal carbon research. Therefore, the remote sensing working group highlighted several of these areas that should be a high priority for the biogeochemical, ocean optics, and remote sensing communities. The needs that were highlighted include:

- Algorithms for 'standard' remotely sensed parameters in the coastal system are rarely validated given the sparse availability of validation data that is coincident with the collected satellite imagery. Given the high-frequency dynamics in coastal systems, this is particularly problematic. The continued maturation of the ocean observatories consisting of long-duration stationary and mobile platforms will greatly expand the available validation data.

- Algorithms are only beginning to become 'carbon-specific' and the errors associated with these empirical algorithms remain an open question. Focused studies on developing and validating satellite-derived carbon products should be pursued. These studies, if combined with other algorithms delineating particle size distributions, and the relative fraction of organic to inorganic particles, would be particularly powerful tools for coastal carbon research efforts.

- Atmospheric corrections are suspect near continents given complex aerosol profiles and the near-continuous presence of haze. The expanding network of ocean observatories may

offer the potential infrastructure to improve our understanding and algorithms.

- Spatial and temporal resolution (days and kilometers) of current observations are barely adequate for coastal process study. Traditionally, revisit intervals by ocean color satellites are at best once per day. The expanding international constellation of ocean color satellites improves this, providing up to 4 to 5 satellite passes a day. This requires very careful inter-calibration between the satellites, which is not a trivial task. Many of these available systems have spatial resolutions down to 350 m and thus will be extremely useful for this coastal carbon effort. Additionally, several efforts are focused on developing geostationary satellites. These systems would greatly benefit this program; however, given current launch schedules they are unlikely to be available within the next five years although strong interest from the coastal carbon community might assist in accelerating the proposed launch dates.

Figure 9.7. 350-m resolution satellite chlorophyll from the Indian OCM satellite.

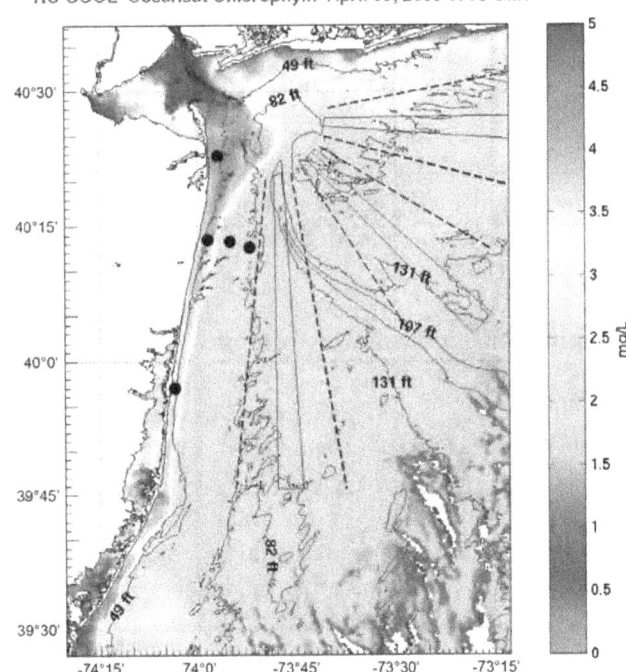

Modeling and synthesis

Predictive models

Without incurring staggering costs, there is probably no combination of observational technology that will cover the four dimensions at adequate resolution and temporal and spatial extent to totally constrain margin carbon-cycling processes. Even in the event that such an observational program were feasible, it would not provide the kind of predictive capability desired for projecting the responses of coastal systems to a changing environment.

These observational shortcomings call for the development of mechanistic models of carbon cycling in coastal margin settings. On continental shelves, carbon cycling is complex, reflecting highly variable atmospheric forcing, strong coupling to the irregular bottom topographies, and buoyant river plumes. The resulting food web dynamics are complicated and are not described in the current generation of global carbon cycle models. Incorporating the actual complexity is not likely to be possible, and our understanding of how detailed the biogeochemical models need to be remains an open question. This is especially true for broad continental shelves where a significant fraction of the organic matter is remineralized and recycled by the food web. Given this, the modeling work group highlighted several areas that need to be addressed to move the community forward.

- Lack of overlap in models at the 10 to 100 km spatial resolution (Figure 9.8) is problematic for resolving many of the coastal processes important for understanding carbon cycles. Even nesting and coupling the diverse number of available models remains a problem, especially as the boundaries where the models are melded are at critical locations. For example, coupling of models often occurs near the continental slope, which then potentially influences how carbon will be exchanged between the continental shelf and deep ocean.

- Many of the unique features of the coastal ocean are still being incorporated into models. For example, sediment resuspension, benthic diagenesis, tides, river-estuarine inputs, and terrestrially forced atmospheric variability need to be described.

- System complexity may overwhelm our ability to interpret model scenarios. For example, how many functional phytoplankton groups need to be incorporated? Should the food web models incorporate descriptions for cellular ballast, size distributions, and multiple grazer communities? Given the likely need for complex models, will there be adequate data for model initialization and validation?

- Data assimilation is still an unproven commodity. This is especially true for food web models. As the data assimilation approaches improve, they will be critically dependent on the availability of adequately temporally and spatially resolved observations.

Historical Data Synthesis

In addition to predictive modeling efforts, study of the continental margins would be greatly aided by the re-examination and synthesis of historical

Figure 9.8. Schematic representation of the development of models at a variety of scales. Models are well developed at large (ocean-basin) and small (watershed) scales, but poorly at the relevant scales for study of the ocean margins.

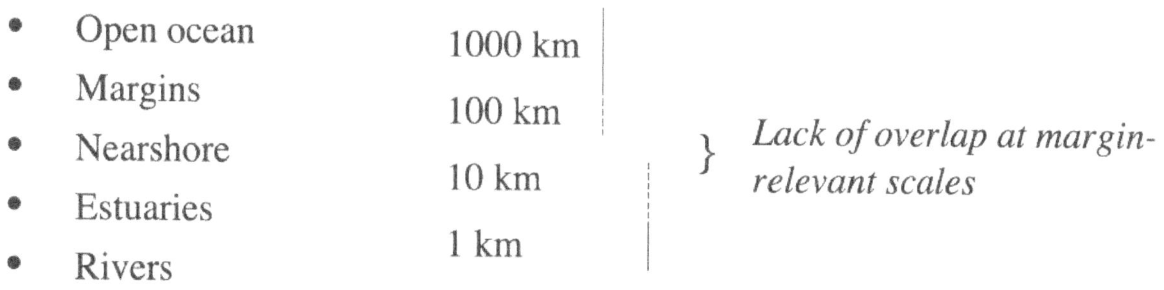

- Open ocean   1000 km
- Margins   100 km
- Nearshore   10 km   } *Lack of overlap at margin-relevant scales*
- Estuaries
- Rivers   1 km

datasets which may not have been fully utilized from the perspective of carbon cycle science. One obvious example is the DOE Ocean Margins Project (OMP), which resulted in an intensive, carbon-focused, process study at a single site on the margin of the Middle Atlantic Bight. While this study was tremendously successful, resulting in a well-subscribed special issue in DSR, funding for the project was terminated before the field data could be fully synthesized. Another example is the LDEO global P_{CO2} database, which includes many measurements off North American coasts that had been excluded from consideration in global flux estimates. These data had been largely ignored prior to the revisitation of this large database by Chavez et al. (2007). Other coastal programs (e.g., CALCOFI, GLOBEC, COOP, OMP) include relevant data (POC, O_2, nutrients, etc.) that have not been fully interpreted in the context of carbon budgets.

There are thought to be large caches of relevant 'gray-literature' data as well. These include data collected for estuarine and nearshore water quality assessments by waterfront municipalities, and ancillary measurements made as part of surveys by fisheries agencies and industry. There is likely to be information gathered by other agencies that may be relevant, such as those governing hydroelectric or mining operations. In addition, the energy extraction industries may have non-proprietary information that could prove useful.

Summary

Oceanography is undergoing a technical evolution that will be critical for the future programs focused on understanding carbon biogeochemistry in the ocean margins. While there is great promise, some major hurdles need to be overcome to realize the full potential of the proposed international observing and modeling networks. For in-water platforms, new small compact sensors are a priority. These sensors will enable the proposed networks. For remote sensing, the development of improved carbon-specific algorithms remains a large challenge for the remote sensing community. Additionally, the potential "gap" of available ocean color satellites will remain a potential problem for future efforts. Efforts such as those outlined in this report could assist in galvanizing community priorities in the coming decade. For the modeling community, fundamental studies of the required complexity for the continental shelf carbon biogeochemistry are required. Model developments remain to be addressed. Also, the development of data techniques for chemical and biological data will necessarily require effort by the community in the coming five years. Despite these challenges, the scientific community has the potential to collect high-resolution data to allow for improved understanding in the near future.

References

Bandstra, L., B. Hales., and T. Takahashi, 2006. High-frequency measurements of total CO_2: Method development and first oceanographic observations. *Marine Chemistry*, 100:24-38.

Berg, P., H. Røy, F. Janssen, V. Meyer, B.B. Jørgensen, M. Huettel, and D. de Beer, 2003. Oxygen uptake by aquatic sediments measured with a novel non-invasive eddy-correlation technique. *Marine Ecology Progress Series*, 261:75-83.

Blackwell, S.M., J. Case, S. Glenn, J. Kohut, M. Moline, M. Purcell, O. Schofield, and C VonAlt, 2002. New AUV Platform for studying near shore bioluminescence structure. In: *Bioluminescence & Chemiluminescence: Progress and Current Applications* [P.E. Stanley and L.J. Kricka (eds.)]. World Scientific, Singapore, pp. 197-200.

Broecker, W.S., and T. Takahashi, 1966. Calcium carbonate precipitation on the Bahama Banks. *Journal of Geophysical Research*, 71:1575-1602.

Byrne, R.H., and J.A. Breland, 1989. High precision multiwavelength pH determinations in seawater using cresol red. *Deep-Sea Research*, 36, 803-810.

Chavez, F.P., T. Takahashi, W.-J. Cai, G. Friederich, B. Hales, R. Wanninkhof, and R. Feely, 2007. Coastal oceans. In: *The First State of the Carbon Cycle Report (SOCCR): The North American Carbon Budget and Implications for the Global Carbon Cycle.* [A.W. King, L. Dilling, G.P. Zimmerman, D.M. Fairman, R.A. Houghton, G. Marland, A.Z. Rose, and T.J. Wilbanks (eds.)]. A report by the U.S. Climate Change Science Program and the Subcommittee on Global Change Research, Washington, DC, pp. 157-166. Available at http://www.climatescience.gov/Library/sap/sap2-2/final-report/default.htm.

DeGrandpre, M.D., T.R. Hammar, S.P. Smith, and F.L. Sayles, 1995. In situ measurements of seawater pCO_2. *Limnology and Oceanography*, 40:969-975.

DeGrandpre, M.D., T.R. Hammar, G.J. Olbu, and C.M. Beatty, 2002. Air-sea CO_2 fluxes on the US Middle Atlantic Bight. *Deep-Sea Research II*, 49:4355-4367.

Eriksen, C.C., T.J. Osse, R.D. Light, T. Wen, T.W. Lehman, P.L. Sabin, J.W. Ballard, and A.M. Chiodi, 2001. Seaglider: A long-range autonomous underwater vehicle for oceanographic research. *IEEE Journal of Oceanic Engineering*, 26:424-436.

Glenn, S., O. Schofield, T.D. Dickey, R. Chant, J. Kohut, H. Barrier, J. Bosch, L. Bowers, E. Creed, C. Haldeman, E. Hunter, J. Kerfoot, C. Mudgal, M. Oliver, H. Roarty, E. Romana, M. Crowley, D. Barrick, and C. Jones, 2004. The expanding role of ocean color and optics in the changing field of operational oceanography. *Oceanography*, 17:86-95.

Glibert, P.M., S. Seitzinger, C.A. Heil, J.M. Burkholder, M.W. Parrow, L.A. Codispoti, and V. Kelly, 2005. The role of eutrophication in the global proliferation of harmful algal blooms: New perspectives and new approaches. *Oceanography*, 18:198-209.

Glud, R.N., M. Kuhl, O. Kohls, and N.B. Ramsing, 1999. Heterogeneity of oxygen production and consumption in a photosynthetic microbial mat as studied by planar optodes. *Journal of Phycology*, 35:270-279.

Hales, B., D. Chipman, and T. Takahashi, 2004a. High-frequency measurement of partial pressure and total concentration of carbon dioxide in seawater using microporous hydrophobic membrane contactors. *Limnology and Oceanography: Methods*, 2:356-364.

Hales, B., T. Takahashi, and L. van Geen, 2004b. High-frequency measurement of seawater chemistry: Flow-injection analysis of macronutrients. *Limnology and Oceanography: Methods*, 2:91-101.

Hales, B., T. Takahashi, and L. Bandstra, 2005a. Atmospheric CO_2 uptake by a coastal upwelling system *Global Biogeochemical Cycles*, 19, doi:10.1029/2004GB002295.

Hales, B., J.N. Moum, P. Covert, and A. Perlin, 2005. Irreversible nitrate fluxes due to turbulent mixing in a coastal upwelling system. *Journal of Geophysical Research*, 110:C10S11, doi:10.1029/2004JC002685.

Hales, B., L. Karp-Boss, A. Perlin, and P. Wheeler, 2006. Oxygen production and carbon sequestration in an upwelling coastal margin. *Global Biogeochemical Cycles*, 20, GB3001, doi:10.1029/2005GB002517.

Holm, C., 2006. *Development of an Autonomous In-Situ Instrument for Long-Term Monitoring of Cu(II) in the Marine Environment*. MS Thesis, Oregon State University.

Johnson, K.S., and L.J. Coletti, 2002. In situ ultraviolet spectrophotometry for high resolution and long-term monitoring of nitrate, bromide and bisulfide in the ocean. *Deep Sea Research I*, 49:1291-1305.

Kaltin, S., C. Haraldsson, and L.G. Anderson, 2005. A rapid method for determination of total dissolved inorganic carbon in seawater with high accuracy and precision. *Marine Chemistry*, 96:53-60.

Moline, M., P.W. Bissett, S. Blackwell, J. Mueller, J. Sevadjian, C. Trees, and R. Zaneveld, 2005. An autonomous vehicle approach for quantifying bioluminescence in ports and harbor. *Proceedings of SPIE* Vol. 5780, doi:10.1117/12.606891.

Reimers, C.E., L.M. Tender, S. Fertig, and W. Wang, 2001. Harvesting energy from the marine sediment-water interface. *Environmental Science and Technology*, 35:192-195

Reimers, C.E., P. Girguis, H.A. Stecher III, L.M. Tender, N. Ryckelynck, and P. Whaling, 2006. Microbial fuel cell energy from an ocean cold seep. *Geobiology*, 4:123.

Roche, M.P., and F.J. Millero, 1998. Measurement of total alkalinity of surface waters using a continuous flowing spectrophotometric technique. *Marine Chemistry*, 60:85-94.

Rudnick, D.L., R.E. Davis, C.C. Eriksen, D.M. Fratantoni, and M.J. Perry, 2004. Underwater gliders for ocean research. *Marine Technology Society Journal*, 38:48-59.

Schofield, O., J. Kohut, D. Aragon, L. Creed, C. Haldeman, J. Kerfoot, H. Roarty, C. Jones, D. Webb, and S.M. Glenn, 2007. Slocum Gliders: Robust and ready. *Journal of Field Robotics*, 24:473-485, doi:10:1009/rob.20200.

Sigleo, A.C., C.W. Mordy, P. Stabeno, and W.E. Frick, 2005. Nitrate variability along the Oregon coast: Estuarine-coastal exchange. *Estuarine, Coastal and Shelf Science*, 64:211-222.

Takahashi, T., S.C. Sutherland, C. Sweeney, A. Poisson, N. Metzl, B. Tilbrook, N. Bates, R. Wanninkhof, R.A. Feely, C. Sabine, J. Olafsson, and Y. Nojiri, 2002. Global sea-air CO_2 flux based on climatological surface ocean pCO_2, and seasonal biological and temperature effects. *Deep Sea Research II*, 49:1601-1622.

Wanninkhof, R., A. Olsen, and J. Triñanes, 2007. Air–sea CO_2 fluxes in the Caribbean Sea from 2002–2004. *Journal of Marine Systems*, 66:272-284.

Weiss, R.F., 1974. Carbon dioxide in water and seawater: The solubility of a non-ideal gas. *Marine Chemistry*, 2:203-215

Wetz, M.S., B. Hales, Z. Chase, P.A. Wheeler, and M.M. Whitney, 2006. Riverine input of macronutrients, iron, and organic matter to the coastal ocean off Oregon, U.S.A., during the winter. *Limnology and Oceanography*, 51:2221-2231.

Wheatcroft, R.A., 1994. Temporal variation in bed configuration and one-dimensional bottom roughness at the mid-shelf STRESS site. *Continental Shelf Research*, 14:1167-1190.

Workshop Conclusions and Recommendations

The workshop presentations and discussions highlighted several general conclusions regarding carbon cycling in the continental margins. These are presented here, followed by a set of recommendations for future carbon cycle research.

- **Coastal carbon cycles are the result of integration of a broad array of processes.**

The air-sea flux of CO_2 is a result of a combination of physical, chemical, biological, and geological processes that must be fully understood before any mechanistic or predictive understanding can be gained. Nowhere else in the ocean are the deep ocean interior, the seafloor, the water column, and the terrestrial biosphere so closely coupled as they are in the coastal ocean. Thus, while extensive coverage of surface pCO_2 measurements in the North American continental margins is a desirable goal that might finally resolve the question of the sign and magnitude of the coastal air-sea CO_2 flux, that alone would only provide limited ability to predict likely changes in carbon cycling in this variable environment in response to likely changes in forcing driven by climate change. As a result, a detailed understanding of the roles that, e.g., circulation, terrestrial inputs, atmospheric (thermal and wind forcing) processes, food webs, benthic processes, and non-C elemental cycling play in determining carbon cycling in a given region will be critical.

- **Coastal systems experience high variability over a wide range of scales.**

Strong air-sea CO_2 flux signals have been reported in all coastal settings, yet constraint of a net flux is extremely challenging due to large spatial and temporal variability on variety of scales. Coastal systems undergo temporal variability that ranges from hours (e.g., due to tides) to days (e.g., due to fluctuations in wind forcing) to months (e.g., due to seasonally variable forcings) to years (e.g., due to ENSO variability) to decades (e.g., due to NAO- and PDO-related fluctuations). Lateral spatial variability can be important at scales ranging from meters (e.g., in estuarine or plume-influenced environments) to kilometers (e.g., in the vicinity of

fronts, filaments, or eddies) to hundreds of kilometers (e.g in the transitions between prevailing wind and ocean current regimes, or the in the influences of seasonal forcing). This is exemplified for surface CO_2 by considering the Chavez et al. (2007) coastally-focused re-examination of the global CO_2 database. This database includes pCO_2 values that range from <100 µatm to >1000 µatm. In some regions, this describes temporal as well as spatial variability. The net flux estimated from this dataset is near zero, but the uncertainties due to the variability in that estimate are large enough to include the largest coastal CO_2 sink (Hales et al. 2005) and source (Cai et al. 2003) estimates, when projected to similar areas. The reasons for these extreme ranges are the variations in the underlying mechanisms that drive the surface CO_2.

- **Coastal systems are dramatically undersampled with respect to carbon cycling.**

The above-described variability would make adequate sampling difficult even with the most ambitious observational efforts; actual efforts have been far from ambitious. Examination of the compilations of Cai et al. (2006) and Borges (2005), where a handful of measurement programs have been taken as representative of the entirety of coastal oceans, illustrate this problem clearly. Even the totality of the nearly 10^6 North American coastal measurements in the global CO_2 database (Chavez et al. 2007) leaves many key regions almost unsampled. The reasons for this undersampling are several. Carbon cycling in general and air-sea CO_2 exchange in particular have not been foci of coastal measurement programs, which have often been more concerned with related issues such as fisheries, eutrophication, hypoxia, etc. As a result, CO_2 measurements have not been routinely made during coastal programs. Coastal CO_2 measurements are sometimes incidentally made as part of large-scale pelagic programs; however, as seen in the distributions of measurements in the Chavez et al. (2007) compilation, measurements made along transit lines of ships of opportunity or on research vessels bound for pelagic sampling efforts do not result in good coverage of the coastal ocean. Even worse, pelagic researchers often 'turn off' measurement systems when crossing Exclusive Economic Zone (EEZ) boundaries from international waters. Traditionally there has been neither routine surface underway measurement of pCO$_2$ made on research vessels operating primarily in the coastal ocean, nor regular deployment of moored CO$_2$ measurement systems on coastal buoys.

- **Coastal ocean regional boundaries are poorly defined**

The definition of the coastal oceans was surprisingly equivocal, as workshop participants debated classifications based on salinity distributions, remotely sensed ocean color, bathymetry, and geographic and geopolitical boundaries. In addition to uncertainty in the boundaries in the cross shelf dimension, it was widely recognized that the simple classification of the North American margins into the six regions describe in Chapters 3 to 8 was overly coarse. Each of those regions consisted of several subregions within which prevalence of processes was distinct.

- **Current measurement techniques need refinement**

Existing measurement technologies, whether in-water ship or mooring-based methods or remote-sensing approaches, have many shortcomings for application in the coastal oceans. The large dynamic ranges over short length and time scales in both parameters of interest and in environmental conditions complicate many measurements. Available in-water measurement platforms include small coastal research vessels and navigational buoys that limit the size of scientific instrumentation that can be deployed. The high cost of measurement systems limits widespread deployment of platform-of-opportunity sensors or analyzers. High variability in measured parameters suggests the need for sensors or analytical systems with larger dynamic ranges and faster responses than has been the case for pelagic study. Corrections for complicating factors, especially with regard to remote sensing, can be significant in coastal settings and are currently poorly defined.

- **Existing coastal datasets have been under-utilized**

The workshop participants also recognized the existence of many datasets that have been under-utilized in the context of coastal carbon-cycle processing. One obvious example is the DOE Ocean Margins Project (OMP), which resulted in an intensive, carbon-focused process study at a single site on the margin of the Middle Atlantic Bight. While this study was tremendously successful, resulting in a well-subscribed special issue in DSR, funding for the project was terminated before the field data could be fully synthesized. Another example is the LDEO global P$_{CO_2}$ database, which includes many measurements off North American coasts that had been excluded from consideration in global flux estimates. These had been largely ignored prior to the revisitation of this large database by Chavez et al. (2007). Other coastal programs (e.g., CALCOFI, GLOBEC, COOP, OMP) include relevant data (POC, O$_2$, nutrients, etc.) that have not been fully interpreted in the context of carbon budgets.

In addition to under-utilized scientific datasets, there are thought to be large caches of relevant 'gray-literature' data. These include data collected for estuarine and nearshore water quality assessments by waterfront municipalities, and ancillary measurements made as part of surveys by fisheries agencies and industry. There is likely to be information gathered by other agencies that may be relevant, such as those governing hydroelectric or mining operations. In addition, the energy extraction industries may have non-proprietary information that could prove useful.

- **Modeling efforts have been limited**

Even though models have many technical shortcomings with respect to their application in the coastal oceans (see previous sections), much has been learned from modeling efforts focused on coastal regions. Unfortunately, most detailed biogeochemical modeling studies have been only local studies of specific regions such as the Oregon coast, the Gulf of Maine, and the southern and central California coast. Carbon cycling in some important large-scale regions, such as the Gulf of Mexico, has never been modeled.

- **Understanding of coastal carbon cycling suffers from several key unknowns**

The major unknowns seen as limiting the understanding of the carbon cycle in coastal waters could be summarized as an ignorance of *net* reactions in the water column and surficial sediments, and *net* fluxes across the important boundaries of the coastal ocean. These include the following:

1. Net community water-column diagenesis:

Despite the many studies showing that f-ratios can vary by an order of magnitude or more in the coastal ocean, even across the shelf at a single location (e.g., Kokkinakis and Wheeler, 1987), gross measurements such as primary productivity in surface waters are still used as benchmarks for comparing the photosynthetic activity in coastal regions (e.g., Walsh, 1988). This measurement is largely irrelevant to carbon cycling questions in relation to a more integrative parameter such as net community productivity.

Little is known about water-column degradation of autochthonous and allochthonous organic carbon, particularly at depths below the euphotic zone. As one example, it is known that rivers transport abundant dissolved and particulate organic matter to the coastal oceans, little of which can be accounted for in the coastal sediments or water column. Is this material simply oxidized in the margins? As another example, it is commonly observed that water column production rates, even in the net sense, are far greater than estimates of sediment carbon degradation and burial (e.g., Hales et al. 2006). Are the products of coastal photosynthesis simply respired at depth in the water column?

Likewise, net reaction of organic matter in sediments of the coastal ocean has been difficult to constrain. The abundant benthic macrofauna, complex sediment structures, compressed redox horizons, and non-diffusive transport in the high-energy environment of the shallow seafloor all complicate traditional measurements and interpretations of net reactions in surficial sediments. The recognition that benthic autotrophy may play a significant role in the net diagenesis in coastal sediments only further complicates matters.

2. Net air-sea exchange of CO_2:

As discussed at length in previous sections, the large dynamic range of variability in surface pCO_2 in the coastal ocean, coupled with the sparseness of measurements in key source/sink regions, has left us in a situation where we cannot say whether the North American coastal oceans represent a significant source or sink of CO_2 from or to the atmosphere, or simply a small integrated annual flux with large local variations.

3. Net chemical fluxes across 'shoreline' boundaries:

Primary river fluxes, while themselves subject to many deficiencies (McKee, 2003), are far better constrained than the ultimate net exchange rates of materials between the continents and the coastal oceans. This exchange can occur across a variety of shorelines, including the mouths of estuaries, the complicated seaward boundaries of structures such as salt marshes or mangrove forests, or through the seafloor itself in the form of groundwater flow. Estuary-ocean exchange might be the simplest of these examples: the site of the exchange is relatively easily defined as the estuary mouth, and the sign and approximate magnitude of the net water delivery rate through the estuary mouth is at least approximated by the primary river flow. Extensive biogeochemical processing within the estuary and active bi-directional tidal exchange through the estuary mouth greatly complicate matters and estuaries can variably act as sinks or sources of elements from the adjacent coast, even for elements that are primarily terrestrially derived.

4. Net exchange between coastal and open ocean:

Net transport of carbon between the coastal and open ocean has been offered as a means to explain the excess of carbon delivery to and production in the coastal ocean (e.g., Hales et al. 2006) over degradation and storage there, and as a means to explain the apparent heterotrophy of the open ocean. These exchanges are usually deduced from imbalances, however, rather than directly observed, and often require complicated transport mechanisms.

Workshop Recommendations

Based on the conclusions and shortcomings identified above, the workshop participants offered several recommendations to guide future carbon cycle research in the North American continental margins.

- **Make coastal carbon cycling research inclusive.**

The first major recommendation from the workshop is that coastal carbon cycle research be defined as inclusively as possible, and that carbon cycling should be broadly included as motivation for a variety of research. For example, researchers studying water transport, or nutrient or trace-metal cycling in coastal settings, would have their efforts encouraged under the umbrella of coastal carbon research; conversely, they would be encouraged to motivate their research by placing it in the context of its impact on carbon cycling, thereby providing data products of direct utility to those more directly focused on carbon cycling. Carbon cycle science is inherently integrative and interdisciplinary, and future research directions should recognize and embrace this fact.

- **Improve coastal carbon cycling observational capabilities.**

The problems with data coverage in the coastal ocean have been discussed at length, and a high priority should be mitigating this problem. The workshop saw three immediate avenues of pursuit in this regard:

1. *Expansion of routine measurements in the coastal setting.*
   All coastal ship and mooring platforms should be considered for instrumentation. This should take the form of modifying existing technologies to be suitable for operation in the coastal ocean. Examples might be:
   - Modification of existing chemical and bio-optical measurement technologies to be smaller, more rugged, less expensive, and have faster response times.
   - Expansion of surface mapping from coastal research vessels and vessels of opportunity.
   - Further instrumentation of existing coastal weather and navigational buoys.

2. *Refinement of remote sensing algorithms.*
   Many improvements might lead to better quantification of relevant remotely observable characteristics of the coastal ocean. Specific areas of emphasis could include:
   - Better atmospheric corrections at continental boundaries.
   - Better resolution between CDOM, chlorophyll, and suspended sediments.
   - Increased spatial and temporal resolution of remote-sensing measurements.
   - Development of algorithms for remote salinity estimation.

3. *Development of new technologies.*
   While some existing measurement technologies may require only minor modifications for automated deployment in the coastal setting (e.g., pCO_2 and pH analysis; O_2, NO_3, and bio-optical sensing), still other relevant measurements are far from this state, including the carbonate system parameters total CO_2 and alkalinity, trace metals, and biological rates. Support should be given to development of:
   - Autonomous analysis systems.
   - In situ sensing technologies.
   - High-resolution automated sampling systems.
   - Low-cost, fast-response, small-footprint systems for deployment on coastal vessel and mooring of opportunity platforms.

- **Synthesize and model existing datasets.**

As discussed previously, there are abundant data relevant to understanding coastal carbon cycling that have either been under-synthesized by, or perhaps not even widely known to, the carbon cycle science community. Included in this lack of analysis is the dearth of ecosystem modeling in the coastal environment. An effort to expand the synthesis and modeling of existing data would serve not only to add value to previous observational programs, but also to guide future work in the field. The broader goals of such work would include definition of subregions in which intensive field programs might be sited; identification of characteristic time and length scales of variability within and across such subregions; identification of the critical processes driving carbon cycling, and identifying likely changes in systems under changing climate regimes. Future observational research programs would greatly benefit from a synthesis and modeling effort that would:

1. *Identify existing data resources from research and monitoring programs, generated by scientific and regulatory agencies and non-governmental organizations (i.e., creating a standardized 'database of databases').*

2. *Develop methods for defining subregions of the coastal ocean around North America.*

3. *Model the carbon cycle in identified subregions to quantify sensitivities to processes and forcings, and develop predictive capabilities.*

4. *Develop predictive algorithms for carbon-relevant parameters from higher resolution data records, including remote sensing.*

- **Objectively define the coastal oceans, and subregions, for targeted carbon cycle study.**

The Workshop participants recommend a definition of the coastal oceans as the Great Lakes plus all aqueous environments bounded at the seaward edge by the exclusive economic zone (EEZ); the shoreward edge by the head of tide; at the upper limit by the sea surface; and at the lower limit by the depth of the shelf break or the base of the sediment mixed layer (whichever was shallower). This definition was chosen for the following reasons: 1) The EEZ essentially always encompasses the North American continent's elevated-biomass 'coastal' satellite signal and the shelf break, and roughly coincides with the shoreward extent of global ocean compilations and measurement programs; 2) the head-of-tide boundary encompasses estuarine systems, and coincides with traditional seaward extent of riverine studies; and 3) the deeper boundary represents an interface through which exchanges with the coastal ocean occur over longer time scales, and encompasses the diagenetically active portion of the sediments where recycling times are short.

Within the above boundaries, the group suggested a further categorization into subregions within which focused study would be representative of the entire subregion. Such a categorization would be expected to take into consideration prevalence of carbon-cycle-relevant processes, terrestrial and atmospheric interactions, and spatial and temporal variability in these, and might be expected to be a preliminary result of synthesis of historical datasets. The group offered a preliminary list of 14 process-defined 'provinces' (Figure 10.1); however these should be seen as suggestive only, and more rigorous definitions should be pursued.

- **Develop a plan for observational study and assimilative modeling of characteristic regimes.**

Following progress in response to the recommendations above, the workshop group recommends a plan for intensive field study and modeling of the identified characteristic subregions. The goal of this work would be to perform an integrated study of the carbon cycle within each

Figure 10.1. Potential boundaries of biogeochemical subregions, within which focused process studies could be executed.

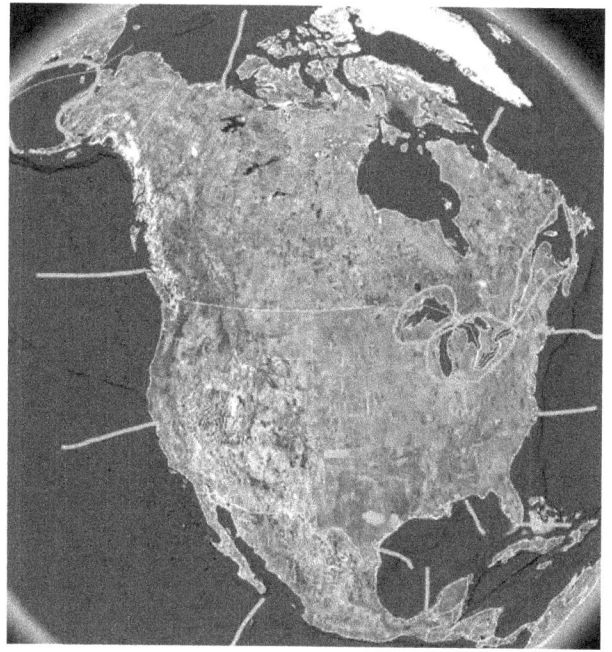

subregion, with emphasis on constraining net fluxes and reactions in the region. Given previous synthesis work defining subregions and scales of variability within each region, the extrapolatability of the results would be known. Such studies would have the following features:

1. *Relevant fluxes, reactions, and processes would be determined separately for each province, following recommendations in this report and results of pre-synthesis studies. Research activities would thus be tailored for each region, rather than following a one-size-fits-all approach.*

2. *Studies would be guided by a control-volume approach with the objective of closing mass balances within a finite space representative of a subregion. The identities and locations of the key boundaries of the control volume would be specific to each region.*

3. *Detailed biogeochemical models of the region would be an integral part of the intensive studies.*

This work would have the following objectives:

1. *Quantification of net carbon-relevant fluxes across key interfaces of the control volume, and net carbon-relevant reactions within the control volume, and detailing the processes that control these.*

2. *Determining relationships between the fluxes, reactions, processes above, and more extensively measurable parameters (e.g., meteorological and satellite data) such that detailed results can be extrapolated to unsampled times and locations of any given region.*

3. *Development of parameterizations of the above such that they can be readily implemented in models.*

4. *Detailed biogeochemical models of the regions would be an integral part of these studies. These would initially guide planning of the fieldwork, and ultimately assimilate the data generated in the field.*

References

Borges, A.V., 2005. Do we have enough pieces of the jigsaw to integrate CO_2 fluxes in the coastal ocean? *Estuaries*, 28:3-27.

Cai, W.-J., Z. Wang, and Y. Wang, 2003. The role of marsh-dominated heterotrophic continental margins in transport of CO_2 between the atmosphere, the land-sea interface and the ocean. *Geophysical Research Letters*, 30:1849.

Cai, W-J., M. Dai, and Y. Wang, 2006. Air-sea exchange of carbon dioxide in ocean margins: A province based synthesis. *Geophysical Research Letters*, 33:L12603, doi:10.1029/2006GL026219.

Chavez, F.P., T. Takahashi, W.-J. Cai, G. Friederich, B. Hales, R. Wanninkhof, and R. Feely, 2007. Coastal oceans. In: *The First State of the Carbon Cycle Report (SOCCR): The North American Carbon Budget and Implications for the Global Carbon Cycle*. [A.W. King, L. Dilling, G.P. Zimmerman, D.M. Fairman, R.A. Houghton, G. Marland, A.Z. Rose, and T.J. Wilbanks (eds.)]. A report by the U.S. Climate Change Science Program and the Subcommittee on Global Change Research, Washington, DC, pp. 157-166. Available at http://www.climatescience.gov/Library/sap/sap2-2/final-report/default.htm.

Hales, B., T. Takahashi, and L. Bandstra, 2005. Atmospheric CO_2 uptake by a coastal upwelling system, *Global Biogeochemical Cycles*, 19, doi:10.1029/2004GB002295.

Hales, B., L. Karp-Boss, A. Perlin, and P. Wheeler, 2006. Oxygen production and carbon sequestration in an upwelling coastal margin. *Global Biogeochemical Cycles*, 20:GB3001, doi:10.1029/2005GB002517.

Kokkinakis, S.A., and P.A. Wheeler, 1987. Nitrogen uptake and phytoplankton growth in coastal upwelling regions. *Limnology and Oceanography*, 32:1112-1123.

McKee, B., 2003. RiOMar: *The Transport, Transformation and Fate of Carbon in River-Dominated Ocean Margins*. Report of the RiOMar Workshop, 1-3 November 2001. Tulane University, New Orleans, LA.

Walsh, J.J., 1988. *On The Nature of Continental Shelves*. Academic Press, 520 pp.